MEGA-URBANIZATION IN THE GLOBAL SOUTH

The global south is entering an 'Urban Age' where, for the first time in history, more people will be living in cities than in the countryside. The logics of this prediction have a dominant framing – rapid urbanization, uncontrolled migration, resource depletion, severe fuel shortages and the breakdown of law and order. We are told that we must be prepared. The solution is simple, they say. Mega-urbanization is an opportunity for economic growth and prosperity. Therefore we must build big, build new and build fast.

With contributions from an international range of established and emerging scholars drawing upon real-world examples, *Mega-Urbanization in the Global South* is the first to use the lens of speed to examine the postcolonial 'urban revolution'. From the mega-urbanization of Lusaka, to the production of satellite cities in Jakarta, to new cities built from scratch in Masdar, Songdo and Rajarhat, this book argues that speed is now the persistent feature of a range of utopian visions that seek to expedite the production of new cities. These 'fast cities' are the enduring images of postcolonial urbanism, which bypass actually existing urbanisms through new power–knowledge coalitions of producing, knowing and governing the city.

The book explores three main themes. Part I examines fast cities as new urban utopias which propagate the illusion that they are 'quick-fix' sustainable solutions to insulate us from future crises. Part II discusses the role of the entrepreneurial state that despite its neoliberalization is playing a key role in shaping mega-urbanization through laws, policies and brute force. Part III finally delves into how fast cities built by entrepreneurial states actually materialize at the scale of regional urbanization rather than as metropolitan growth. This book explores the contradictions between intended and unintended outcomes of fast cities and points to their fault lines between state sovereignty, capital accumulation and citizenship. It concludes with a vision and manifesto for 'slow' and decelerated urbanism.

This timely and original book presents urban scholars with the theoretical, empirical and methodological challenges of mega-urbanization in the global south, as well as highlighting new theoretical agendas and empirical analyses that these new forms of city-making bring to the fore.

Ayona Datta is Reader in Human Geography at King's College London, UK.

Abdul Shaban is Professor at the School of Development Studies and Deputy Director (Tuljapur campus), Tata Institute of Social Sciences, Mumbai, India.

Routledge Studies in Urbanism and the City

This series offers a forum for original and innovative research that engages with key debates and concepts in the field. Titles within the series range from empirical investigations to theoretical engagements, offering international perspectives and multidisciplinary dialogues across the social sciences and humanities, from urban studies, planning, geography, geohumanities, sociology, politics, the arts, cultural studies, philosophy and literature.

For a full list of titles in this series, please visit www.routledge.com/series/RSUC

MEGA-URBANIZATION IN THE GLOBAL SOUTH

Fast cities and new urban utopias
of the postcolonial state

*Edited by Ayona Datta and
Abdul Shaban*

Routledge
Taylor & Francis Group

LONDON AND NEW YORK

First published 2017 by Routledge

2 Park Square, Milton Park, Abingdon, Oxfordshire OX14 4RN
52 Vanderbilt Avenue, New York, NY 10017

Routledge is an imprint of the Taylor & Francis Group, an informa business

First issued in paperback 2020

British Library Cataloguing in Publication Data
A catalogue record for this book is available from the British Library

Library of Congress Cataloging in Publication Data
Names: Datta, Ayona, editor. | Shaban, Abdul, editor.
Title: Mega-urbanization in the global South : fast cities and new urban utopias of the postcolonial state / edited by Ayona Datta and Abdul Shaban.
Description: New York, NY : Routledge, 2016. | Series: Routledge studies in urbanism and the city | Includes bibliographical references and index.
Identifiers: LCCN 2016025153| ISBN 9780415745512
    (hardback : alk. paper) | ISBN 9781315797830 (ebook)
Subjects: LCSH: Urbanization—Developing countries—Case studies. |
    City planning—Developing countries—Case studies. |
    Urban policy-Developing countries—Case studies.
Classification: LCC HT384.D44 M44 2016 | DDC 307.1/216091724—dc23
LC record available at https://lccn.loc.gov/2016025153

ISBN: 978-0-415-74551-2 (hbk)
ISBN: 978-0-367-59581-4 (pbk)

Typeset in Sabon Std
by Swales & Willis Ltd, Exeter, Devon, UK

CONTENTS

ILLUSTRATIONS

Figures

Tables

CONTRIBUTORS

Federico Cugurullo is Assistant Professor in Smart and Sustainable Urbanism at Trinity College Dublin. His research is positioned at the intersection of urban geography and political philosophy, and explores how ideas of sustainability are cultivated and implemented across geographical spaces, with a focus on projects for eco-cities and smart cities.

Ayona Datta is Reader in Human Geography at King's College London. Her research and writing use approaches from sociology, anthropology, and feminist and critical geography, and broadly focuses on the politics of citizenship and urbanization across the global north and south. She is co-editor of *Translocal Geographies: Spaces, places, connections* (2011) and author of *The Illegal City: Space, law and gender in a Delhi squatter settlement* (2012) published by Ashgate. She is working on another co-edited book (contract with Zed) on 'Ecological citizenships in the global south', due in 2017. Ayona is the author of over 30 articles in peer-reviewed journals as well as producer/director of two films, 'City bypassed' and 'City forgotten'. Ayona is editor of two journals – *Urban Geography* and *Dialogues in Human Geography* – and she sits on the editorial boards of the journals *ACME*, *Antipode*, *Gender, Place and Culture* and *Society and Space*.

Tommy Firman is Professor in the School of Architecture, Planning, and Policy Development (SAPPK) at the Institute of Technology, Bandung (ITB), Indonesia. Dr Firman is former Chair of the Department of Regional and City Planning; Dean of Faculty of Civil Engineering and Planning; Director (Dean) of Graduate School; and Chair of The Faculty Senate, at the Institute. He also served as a member of the Indonesian National Research Council (DRN) from 1999 to 2004. Dr Firman has written numerous book chapters and peer-reviewed articles, on urbanization and urban and regional development planning in Indonesia, for several international journals. He has also been a member of the international advisory board of several international journals, including *Habitat International*, *International Development Planning Review* and *Asian Population Studies*. Dr Firman is a recipient of the Otto Koenigsberger Prize from Elsevier Publishers for *Habitat International* (2000). In 1991 he received the 'Ristek' Award from the Ministry of Research and Technology, and the Hantaru (National Spatial Planning Day) Award from the Ministry of Land Affairs and Spatial Planning, Republic of Indonesia, in 2015.

Delik Hudalah is Associate Professor at the School of Architecture, Planning and Policy Development and Senior Fellow at the Infrastructure and Regional Development Research Center, Bandung Institute of Technology (ITB). He holds a PhD in planning (2010) and was a post-doctoral research fellow (2012) at the Faculty of Spatial Sciences, University of Groningen. His particular interests are in the interfaces between urban and rural changes, between socio-economic development and environmental protection, and between global forces and local politics in the production of edge urban spaces, which have become key issues in the planning of emerging Asian metropoles and mega-urban regions. He has participated in various research projects on, among other things, post-suburbanization and industrial deconcentration in Greater Jakarta, peri-urban land use change in European city-regions, and metropolitan governance in Indonesia.

Ratoola Kundu is Assistant Professor in the School of Habitat Studies at Tata Institute of Social Sciences (TISS), Mumbai, where she teaches urban transport policy, planning and practice, urban planning theory and practice, planning and the Indian city, and urban studio and research design. Her research interests include planning and its intersections with urban informality, the processes and impacts of peri-urbanization, and the inclusion and exclusion of groups within and through plans and policies. She has been involved in collaborative projects on the planning and development of small and medium towns at the Centre for Urban Policy and Governance at TISS. She is currently examining the contested processes of social and spatial transformations within one of Mumbai's oldest inner-city neighbourhoods, Kamathipura, with a focus on the precarity of livelihoods, particularly those of very poor and marginalized migrants and commercial sex workers.

Matthew Lane is a PhD student at Lancaster University, UK. His research focuses on the development of sustainable urban design policies for cities in different regions of the world, with previous fieldwork focused on the United States and Zambia. Coming from an urban and economic geography background, his interests focus on the role of power in what is defined as the 'relational production' of urban planning for the interconnected cities of the world. In addition, Matt has also been involved in other related research in and around Lusaka, Zambia, where he is currently based. This includes topics such as the criminalization of informal street-food vendors and the impact of foreign NGOs on existent structures of informality.

Braulio Eduardo Morera is an architect and urban designer. He divides his time between practice with the 100 Resilient Cities programme, supported by the Rockefeller Foundation, and a PhD at the Urban

Lab, University College London. He trained at the Pontifical Catholic University of Chile and obtained an MSc in City Design and Social Science at the London School of Economics. His professional experience includes major urban plans such as Dongtan Eco-city and Wanzhuang Eco-city in China, and a number of urban projects in Malaysia, Singapore, Saudi Arabia, Spain and the UK. Drawing on case studies in the Americas and Asia, his current practice focuses on concepts, policies and measurement of city resilience. His research interests revolve around the politics associated with the communication of urban ideas such as sustainability, resilience and smart cities. With a strong emphasis on Chinese case studies, his recent publications explore ideas of representations of nature and the notion of the urban image.

Martin J. Murray is Professor of Urban Planning at Taubman College of Architecture and Urban Planning, University of Michigan (Ann Arbor). His work engages the fields of urban studies and planning, cultural geography, distressed urbanism, development, historical sociology, African studies, and global urbanism. Professor Murray is author of two books (and one in preparation) on city building and spatial politics in Johannesburg after apartheid. The first – *Taming the Disorderly City* (Cornell University Press, 2008) – examines the challenges for urban planning in Johannesburg after the end of apartheid. The second – *City of Extremes: Spatial Politics in Johannesburg* (Duke University Press, 2011) – looks at the spatiality of wealth and poverty in Johannesburg. The third manuscript in preparation – *Panic City: Johannesburg in the Popular Imagination* – investigates the intersection of public policing and private security in contemporary Johannesburg. He has also published *Commemorating and Forgetting: Challenges for a New South Africa* (University of Minnesota Press, 2013). His most recent book is (tentatively) titled *Urbanism of Exception: City Building in the 21st Century* (Cambridge University Press, forthcoming). His ongoing research focuses on questions related to holistic, master-planned 'private cities' currently under construction or in the planning stages in in urban Africa.

Agatino Rizzo is Associate Professor of Urban Planning and Design in the Architecture Research Group at Luleå University of Technology in Sweden. He has published a number of articles on rapid urbanization, planning and emerging urban regions in South East Asia and the Arab Gulf Region. Between 2010 and 2013, he taught at Qatar University and was lead investigator of a three-year research project on sustainable urban development in greater Doha funded by the Qatar National Research Fund. His main research interests include topics such as comparative/relational urbanism, consumption and knowledge megaprojects, and urban growth/shrinkage in developing and developed countries.

Abdul Shaban is Professor at the School of Development Studies and Deputy Director (Tuljapur campus), Tata Institute of Social Sciences, Mumbai. He is author of *Mumbai: Political Economy of Crime and Space* (Orient Blackswan, 2010) and editor of *Lives of Muslims in India: Politics, Exclusion and Violence* (Routledge, 2012) and *Muslims in Urban India: Development and Exclusion* (Concept Publishing, New Delhi, 2013). He has published several papers on development and deprivation in India, and the production of ethnic spaces in cities. He was a member of the Study Group appointed by the Government of Maharashtra to assess the 'Social, Economic and Educational Status of Muslims in Maharashtra' (2012–13) and a member of the 'Post-Sachar Evaluation Committee' appointed by the Ministry of Minority Affairs, Government of India, New Delhi, to assess the implementation of the programmes for the development of Muslims and other religious minorities. He has authored several reports for the Maharashtra State Minorities Commission, Government of Maharashtra, Government of India, World Bank, and national and international corporate groups.

Hyun Bang Shin is Associate Professor of Geography and Urban Studies in the Department of Geography and Environment at the London School of Economics and Political Science. His research focuses on the critical analysis of the political economic dynamics of urbanization and urban development, with particular attention to Asian cities. He writes on speculative urbanization, the politics of redevelopment and displacement, gentrification, and urban spectacles. His recent books include a co-edited volume, *Global Gentrifications: Uneven Development and Displacement* (Policy Press, 2015), and a monograph, *Planetary Gentrification* (Polity Press, 2016). Hyun is currently working on two other book projects, including a monograph, *Making China Urban* (Routledge), and a co-edited volume, *Contesting Urban Space in East Asia* (Palgrave Macmillan). He serves the journal *CITY* as a senior editor, is a board member (trustee) of the Urban Studies Foundation, and sits on the international advisory board of the journal *Antipode* and on the editorial board of the journals *City, Culture and Society* and *China City Planning Review*. He is also an organizer of the Urban Salon, an interdisciplinary London-based seminar series.

Vanessa Watson is Professor of City Planning in the School of Architecture, Planning and Geomatics at the University of Cape Town, South Africa. She chairs the executive committee of the African Centre for Cities and is a Fellow of the University of Cape Town. She holds degrees from the Universities of Natal, Cape Town and the Architectural Association of London, and a PhD from the University of Witwatersrand. She is the new Global South Editor for *Urban Studies* and is on the editorial boards of *Planning Theory*, *Planning Practice and Research*, the *Journal of Planning Education and Research* and *Progress in Planning*.

ACKNOWLEDGEMENTS

The idea for this book was sown by us in the small office Ayona held during her time as Lecturer at the London School of Economics (LSE). Shaban had come to the LSE as Commonwealth Research Fellow to work with Ayona on a project idea on India's new cities. We spent several afternoons discussing possible themes on this topic and after several cups of coffee decided to work on the notion of private cities. That was the prequel to 'Fast Cities'. We are therefore grateful to the Commonwealth Foundation and to the LSE for providing the funding and space for initializing this conversation.

This book has taken several years to materialize, notwithstanding our own personal and professional responsibilities which always kept pushing this project to the backburner. We are grateful to the earlier editor Faye Leerink for her support in helping us put the book idea together and taking us through contract. We remain forever grateful to Egle Zigaite, editorial assistant, for her guidance thereafter, and more importantly for showing immense patience through the multiple delays in delivering the manuscript of this book.

The Introduction to this book has benefitted from several questions and discussions we had with audiences during invited talks and presentations by Ayona at the University of Oxford; University of Antwerp; MakeCity Festival, Berlin; University of Strathclyde; Open University; University of York.; Eichstatt-Ingolstadt University; National University of Ireland, Maynooth; Humboldt University, Berlin; COMPAS, Oxford University; and several others. Parts of the Introduction are adapted from the paper written by Ayona titled 'New Urban Utopias of Postcolonial India: Entrepreneurial urbanization in Dholera smart city, Gujarat', published in *Dialogues in Human Geography*, vol. 5(1), 3–22. Ayona is grateful to Rohit Madan for his invaluable help in keeping everything else under control while she was finishing the book.

Shaban is thankful to Professor S Parasuraman, Director at the Tata Institute of Social Sciences, for his support and encouragement. We also thank Ujjwal Dadhich for his help in keying in the corrections in the concluding chapter. Shaban also acknowledges with thanks Sanjukta Sattar, Rahil Shaban, Professor Bipin Das and Suraiya K for their invaluable help in various ways.

1

INTRODUCTION

Fast cities in an urban age

Ayona Datta

The world is entering an 'urban age', it seems. We are continuously hearing doomsday predictions about the impending global urban crisis where, for the first time in the history of humankind, more people will be living in cities than in the countryside. The global south is apparently entering an urban age at a faster rate than the global north. We continuously see impressive graphs, pie charts and simulations of this impending urban age presented by 'experts' from global consultancies. The logics of these predictions have a dominant framing – rapid urbanization, uncontrolled migration, resource depletion, severe fuel shortages, and the breakdown of law and order. We are told that megacities such as Mumbai, Johannesburg, Jakarta and others face severe urban crises in the near future. So, we must be prepared. The solution is simple, they say. We should see urbanization as an opportunity and not as a challenge. We must build new cities to reverse the doomsday predictions. And we must build these cities *fast*.

The notion of 'urban crisis' is not new to this generation. It can be argued that western cities have faced different forms of crisis in each decade. From industrial pollution in the 19th century to urban protests in the 1960s, peak-oil shortages in the 1970s, white-flight and inner-city decline in the 1980s, terrorism since 9/11 and more recently economic austerity since the financial crash, cities have been at the centre of diverse geopolitical crises in economics, culture and society. Consequently, cities have also been at the centre of some radical political, economic and planning 'solutions' to these crises, which attempt to reconceptualize the relationship between cities and nature, technology, culture, and society at large. A number of city tropes (such as garden city, radiant city, sustainable city, intelligent city, eco-city and smart city) have served to discursively, visually and politically sustain the utopian idea that urban planning can provide solutions to a range of social and economic crises (Datta 2012, Caprotti 2015). Building new cities under these tropes remains one of the most popular features of modern times.

Right now, across the global south, a 'world of new cities' (Moser et al. 2015, 74) is being conceived and built at an unprecedented pace. China and India have been leading this 'urban revolution' with hundreds of new cities under construction or at blueprint stage. In China, Dongtan remains one of the most cited new 'sustainable' cities of current times. In India, large-scale programmes of public–private partnerships in building new cities (such as Lavasa, GIFT, Dholera, Palava, Rajarhat and so on) has already captured the aspirations of its urban population. Across South-East Asia new cities have acquired such momentum that a 'New Cities Summit 2015' was hosted in Jakarta to discuss how 'seizing the urban moment' could locate cities 'at the heart of growth and development' (New Cities Summit 2015). Even cities in the so-called 'fourth world economies' (Shatkin 2007) such as Cambodia and Vietnam are showing increased engagement with new city-making. Across the Middle-East a number of new cities under construction have already begun to emerge as prototypes of a global urban future. For example, in the UAE, Masdar is repeatedly cited as an example of a sustainable, ecologically conscious smart city of the future. King Abdullah Economic City in Saudi Arabia, Qatar Knowledge City and Khabary Future City in Qatar, and Al-Irfan in Oman are just some of the examples through which Middle-Eastern countries are claiming to enter a post-carbon age. Across Africa, a continent that is arguably a 'new' entrant to the Urban Age, one that has so far been characterized by 'rogue urbanism' (Pieterse and Simone 2013), different countries are now aligning themselves clearly with urban growth economies. Eko Atlantic in Nigeria, Hope City in Accra and Modderfontein in South Africa are some of the cities under construction that are emerging as the new face of urban Africa. These new cities are characterized by a commitment to 'building from scratch' (Herbert and Murray 2015) and promoting a form of 'privatized urbanism' and spatial restructuring in city regions.

These new cities use the rhetorics of 'crisis of urbanization', the 'impending Urban Age', 'sustainable futures' and several other prophesies to highlight the urgency of their need. They are represented through impressive simulated walkthroughs, interactive maps, charts and graphs. They are conceived at a scale and speed unprecedented in modern times. They are part of massive masterplanning and mega-urbanization strategies of emerging economies. They incorporate all the modern features and amenities of global lifestyles, as well as the technology of physical and virtual connectivity for their future residents. These new cities are the focus of investigation in this book.

What is 'fast' in these new cities of the Urban Age? Why should we pay attention to the resurgence of new city-making across the global south? How are these new cities different from the earlier modernist city-making initiatives? How are they connected to the history of postcolonial urban planning and how do they project themselves as urban futures? In short, what is to be gained by examining these new cities?

In this book, we argue that speed is now the persistent feature of new city-making as a way out of crisis in the global south. Speed continues to be a prerequisite to conceptualizing and legitimizing these cities as 'solutions' to the crises of urbanization, migration and climate change. Speed builds upon the rhetorics of urgency but takes it further in producing a range of visions, imageries and fantasies of time–space compression that expedite the circulation of global capital and its materialization into new cities in different regions. It is through their claims to a speed of knowing and governing that these new cities aim to bypass the seeming 'failures' of existing megacities unable to cope with rising crime, pollution and population. It is through the speed of a global transfer of skills, technology and expert knowledge that these new cities strive to achieve their 'potential' as sustainable cities of the future. And it is by constructing earlier urbanization paradigms as 'slow' that the new cities claim to 'seize' the urban moment, 'jumpstart' their economies and 'leapfrog' into sustainable urban futures. The constructed binaries of speed and inertia now dominate the legitimacy of new cities as national 'economic priorities' across global south countries. We call these 'fast cities'.

The terminology of 'fast cities' has so far been used to refer to rapidly growing cities – in other words, economically booming cities. Such cities are seen to be characterized by innovation, entrepreneurship and growth. Their description as 'fast' so far refers to their expansion as centres of commerce or economy and consequently their physical expansion beyond their pre-existing boundaries. A recent report in the US (Scott 2015) labelled New York, Seattle, Dallas and so on as fast cities since these were some of the top 10 'capitals of entrepreneurship'. Another report from Canada (Pembina Institute nd) characterized fast cities as those with the most efficient rapid transit such as Toronto, Montreal, Vancouver, Calgary and Ottawa. 'Fast' cities have also for some time been synonymous with 'start-up' businesses that lead to rapid economic and spatial growth. Fast cities are arguably the success stories of a global neoliberal urbanism.

The 'fast cities' we examine in this book share similar features with those above. Yet they sit uncomfortably between neoliberal and postcolonial narratives of the urban revolution. Despite the highly celebratory nature of the connections between speed, growth and urbanization made by these new cities, urban studies has so far been slow in examining the role of new city-making in the recasting of postcolonial urbanism. The repetition and diversity of new cities across emerging economies indicate a new notion of 'regulated time, governed by rational laws, but in contact with what is least rational in human beings: the lived, the carnal, the body' (Lefebvre 2004, 9). Fast cities are produced from the geopolitical trajectory of a 'city-in-a-box' (Lindsay 2011), driven by 'fast policy' (Peck 2002) and 'expert' knowledge exchange across the world. They acquire global exchange value as the tropes of new urbanism, sustainable city, eco-city or smart city circulate across continents and countries, and metamorphose into new urban forms that embody the regional interpretations of a global neoliberal urbanism. In doing so, these new cities intensify and accelerate urbanization of city regions, articulating historical, social and political capabilities of the local by 'mutating' (Rapoport 2014) from their universal global form. Meanwhile, these new cities could be argued to be experiments in 'worlding' (Roy and Ong 2011) akin to a 'Dubaisation of Africa' (Choplin and Franck 2010), referencing other new cities such as Masdar and Songdo in the global south in attempts to rival their economic growth stories. In this mode of 'assemblage urbanism' (McFarlane 2011), cities in the global south are no longer held hostage to the transfer of policy, skills and knowledge from the west. Rather, new cities induce a frenetic urbanization that attempts to break away from earlier colonial forms of urbanization and masterplanning and establish a new postcolonial identity freed from the past. Yet this referencing often encounters the gritty realities of negotiating planning laws, land acquisition and local resistance when attempts are made to materialize them. When the speed of policy mobility, skills transfer and knowledge exchange collide with the regional and national processes of building the new city, 'speed' often turns into 'inertia'. Postcolonial states then attempt to bypass inertia through new power coalitions of knowing and governing the city. In doing so, new cities recast historic colonial and postcolonial social divisions into a 21st-century mould of fast urbanism.

We argue that understanding these new cities requires an analytical lens of speed, time and scale that has so far been less evident in urban and postcolonial studies. For Lefebvre (2004), any analysis of urbanization must take into consideration the cyclical and linear notion of time. This has important implications for fast cities. For a start, new cities as a solution to 'urban crisis' highlight the marking, measuring and spatializing of a regulated and linear time. Global rhythms of urban crises also have their own measures of time – frequency, consistency, predictions, action, outcomes. These are predicated on the relative construction of speed – of urbanization, migration and climate change. Second, we suggest that fast cities re-open postcolonial insecurities around modernity, development and poverty, to the imperatives of speed. We see fast cities as articulating new state-expert power coalitions around speed that had earlier been indistinguishable from the dominant narratives of development and modernity in postcolonial urbanism. Fast cities present a particular coming together of what Roy (2011, 307) describes as 'speed, hysteria, mass dreams' in postcolonial urbanism. In other words, fast cities present a re-emergence of the postcolonial state desirous of distinction, differentiation and disentanglement from the colonial burden – a reinvention through new utopian imaginings of the city. Thus although they signal new urban futures, the fast cities we examine in this book are rooted in anxieties of postcolonial subjectivity that underline the vulnerabilities of the present. In doing so they become what Grosz (2001, 49) argues is more of a measure of the 'status and permeability of the present than they are indices of transformation or guarantees of a present-to-be'.

A challenge for global urban studies?

Most of the new cities we examine in this book are still present on the drawing board or in construction sites. Measuring their 'success' or 'failure' remains a theoretical, empirical and methodological challenge. As cities without existing economic indicators, they cannot be measured for the fulfilment of the grandiose claims of tackling economic growth or rural migration. As new cities built from scratch, they cannot be measured through the actually existing models of governance. And as cities without citizens, they cannot be studied through the ethnographies of everyday urban life that characterize the rich urban sociologies of megacities of the global south. Nevertheless, it is clear from the handful of the initial phases of these cities built so far (mainly Masdar and Songdo) that they are struggling to achieve their target population and therefore failing to uphold the elaborate mythmaking around new city-building.

Why write an entire book on new cities, when they are but only a fraction of the urban development projects in the global south? How can we examine these new cities when they are still largely present in national growth policies, on the drawing boards of planners, on the webpages of IT consultancies, in glossy reports of growth coalitions, and in the desires and aspirations of citizens? How can we study a 'thing' that has not yet fully materialized, been lived in? How do we imagine their contribution to the combined urban futures in the global south?

In this book, we examine fast cities as a temporal moment in the spatialization of the current global crises. Focusing on this moment does not mean a rejection of history or continuity, but rather a way of thinking through postcolonial urbanism as a mode of succession of the forces of the past by outlandish claims to utopian urban futures. Although widely diverse in their temporalities of capital, scale and space, the rise of fast cities in the global south is a barometer of urban aspirations in postcolonial contexts. While acknowledging then that there are key differences in the historical, political, social and cultural processes through which these cities are marketed, materialized and inhabited, there are also important elements of comparison between them. A comparative gesture across these cities, then, begins with challenging hierarchies and permanencies and sees each of them as a form of geopolitically relevant temporality, a modality of becoming in local/regional contexts through distinct socio-political, cultural and economic time-spaces.

We argue, first, that any study of fast cities needs to take account of the transformation in the notion of speed, time and duration across spaces, places and scales of knowledge transfer, skills generation, policy mobility, conceptualization, implementation and governance. Thus, any examination of the accelerated growth of new cities in recent years has to take into consideration the experiments with speed in regional histories. For example, Chandigarh was constructed within a span of 10 years and Brasilia was built within a span of 41 months, both with a view to accommodating rapid industrialization and rural–urban migration. More recently, in the 1980s, the speed of construction of the first tallest building in China – the Shenzhen International Trade Center – earned the town the name 'Shenzhen speed' (Fen et al. 2016). Its rapid economic growth served to sustain this label and reinforce the myth that building big and fast was the route to economic growth. Shenzhen was also one of the key cities that inspired the Indian Prime Minister Narendra Modi who modelled one of India's 'first' smart cities, Gujarat Industrial and Financial-Tech (GIFT), on its image – a city whose construction is to be completed in a span of five years. Understanding how a transformation in approaches to speed, time and scale in urban development is linked to the crisis of urbanization therefore becomes key to examining fast cities.

We are interested here in time and speed as 'the time before time, the time of the interval, the time of non-time, . . . the "fate" of space' (Grosz 2001, 111) and by extension, the fate of postcolonial urbanism. We examine new cities as a mode of temporality that is striving continuously to establish a universal language for the future, to stand for postcolonial urbanism as a rejection of its historical connection to colonial urban planning. We under-stand fast cities as a form of postcoloniality that is deeply counter-historical, or in other words, a performative strategy refusing to embody the burden of the past. Postcoloniality in this context acknowledges history if only to self-consciously highlight its difference and discontinuity from it. In doing so, fast cities present a contestation between temporality and timelessness. In attempts to 'see from the south' (Watson 2009) when seeking a universal solution to urban crises, fast cities claim to seize and control future time and thereby our shared urban futures.

Second, we argue in this book that while fast cities claim to provide 'planned' solutions to the crisis of urbanization, they can no longer be examined through conventional methods of capturing economic indica-tors, urban agglomerations or rural–urban migration trends. Rather they need to be examined as city models that scale up to regional urbanization. While they might be seen to repeat earlier urbanization patterns of 'edge cities' (Garreau 1992) or 'new towns' (Hogan and Houston 2002), they now primarily promote new forms of 'speculative urbanization' (Goldman 2011) at a rapid pace and on a grand scale. While the earlier building of new towns, industrial cities, knowledge cities, satellite cities and so on went hand in hand with a logic of 'metropolitanism' or the expansion of existing cities for 'planned growth', new cities now advocate a 'bypass urbanization' (Bhattacharya and Sanyal 2011) that circumvents the challenges of actually existing urbanism in megacities to channel funding, resources and services away from existing cities.

In the global south, urban and regional planning are now increasingly driven by the logics of mega-cluster city regions to encourage the process of corridor or cluster urbanization. McKinsey and other global consultancies now aggressively promote the creation of city mega-clusters to respond to urbanization as an 'opportunity' for economic growth. This form of urbanization relies upon large-scale commoditization of land along mega-infrastructure economic corridors, transforming 'unproductive' commons into 'valuable' real estate. Indeed, as Nair (2015) notes in her research on the Hyderabad-Bangalore economic corridor, megacities remain a mere reference point, while new cities along the corridor are the focus of development, planning and investment. In their strategic location along these corridors, fast cities underline the extension of earlier forms of urban segregation to new forms of expulsions from land, livelihoods and resources (Sassen 2014). For example, the recent 'Make in India' national programme to build at least 25 new mega-city nodes along industrial and economic corridors in the next decade highlights the speed, time and scale of imagining new cities in sync with urbanization strategies. As part of a shift in development paradigms in the global south (in China, Malaysia, Korea, Brazil and other countries) from removing urban informality to constructing new cities (Moser 2010, Percival and Waley 2012, Watson 2013), fast cities reflect how 'utopian imaginings' (Bunnell and Das 2010) around big, bold innovative approaches to economic growth have become central to contemporary urbanization in the global south. Any analysis of these new cities must therefore understand them as key drivers of an ideology of accelerated regional urbanization.

Third, although fast cities can be studied in a context of a 'global privatisation of urban space' (Hogan et al. 2012), they are much more state-driven initiatives than earlier acknowledged. Fast cities can be seen as an extension of neoliberal urban development in Asia, Africa, the Middle-East and Latin America, which share several features with private gated communities seen in the post-liberalization phase of these countries. They are arguably scaled-up versions of gated communities, and other spatial manifestations of built form in the global south built through private-sector-led development in the context of a rapidly neoliberalizing postcolonial state. These gated communities have variously been called 'private cities' (Glasze et al. 2006), 'enterprise cities' (MLTM 2005), 'new enclaves' (Atkinson and Blandy 2005), 'privato-polis' (Shatkin 2011), 'enclave urbanism' (Breitung 2012) and even 'spaces of global cultures' (King 2004). This scaling up from the logics of private residential gated development to new 'private' cities has been largely led by global private sector investment in urban real estate. In these new cities, urban planning, development, governance, and control of law and order are in the hands of the private sector that often serve the interests and aspirations of the political elite and middle classes (Choe et al. 2008), producing a 'new urban colonialism' (Atkinson and Blandy 2005) at a rapid pace and scale.

The rise of new cities, however, can fit neither the outlines of 'global gentrifications' (Lees et al. 2015) nor the debates on neoliberalization of the postcolonial state. This is because private-sector-led planning does not mean the absence of state, but rather a restructuring of laws, policies and practices by the state which uses its executive powers to drive urbanization (Kundu 2014). While it might seem that the rise of these new 'city states' competing for investment, innovation and growth produce a world of 'entrepreneurial cities' (Harvey 1989), it is in actuality the 'urbanization of the state' (Hsing 2012) that actively induces the transformation and control of territories in the service of fast cities. This is a new occurrence that is different from earlier 'company towns' that were historically associated with the altruistic vision of industrialists to provide housing and wellbeing for their workers close to centres of production. While some of these early private cities (such as Jamshedpur) still continue to be governed and managed by private companies, they remain muted in speed and scale. The new cities in India, on the other hand, have been aggressively promoted through the laws, regulations, policies and visual representations of an 'entrepreneurial state' (Mazzuccato 2013).

Despite the shrinking of this state from public investments and a neoliberalization of its development initiatives, the entrepreneurial state now more aggressively institutes sovereign rule through the construction of new cities. While 'private urbanism' in itself is not new, rapid urbanization in the global south driven by the increasing desire of the entrepreneurial state to control and govern territory (Schindler 2015) is certainly a 'new' feature of postcolonial urbanism that is gaining increased precedence in the emerging economies. To this end, the entrepreneurial state employs a range of knowledge coalitions or 'global intelligence Corps' (Olds 2001) backed up by 'agents of the state' (Peck 2002) to legitimize its aspirations and mobilize public opinion among the rising middle classes to support these initiatives. The challenge then is to understand the 'global privatisation of urban space' (Hogan and Houston 2002) in this world of new cities without using the template of 'southern' gentrifications (Ghertner 2014). The challenge is also to provide a 'thick' description of the processes of 'enclaving' and 'accumulation by dispossession' through the articulation of 'statecraft' (Theodore and Peck 2015) in producing a range of new cities across the global south.

In this book we examine the ascendance of fast cities through the politics of speed and urgency. In taking up the challenge of studying fast cities, we align ourselves with Chakrabarty's (2000, 34) suggestion of 'developing the problematic of non-metropolitan histories' by unpacking and making visible the 'repression and violence that are as instrumental' in the valorization of fast cities 'as is the persuasive power of its rhetorical strategies' (44). Examining these cities means not just 'identifying and empowering a new loci of enunciation' (Sheppard et al. 2013, 895) that situates them in their historical, regional and temporal contexts, but also unpacking the 'ambivalences, contradictions, the use of force, and the tragedies and ironies' (Chakrabarty 2000, 43) associated with their claims to urbanization as a route out of crises. While the rhetorics and representations of fast cities have been about the global circulation of 'win-win solutions', it appears very different if we refocus our attention on 'local history, and a view of urban change not as imposed from above but rather as an inherently negotiated process' from below (Shaktin 2007, 6). Examining the fast city means identifying and unpacking the temporal nature of its utopian claims that are rooted in its colonial and postcolonial histories, its association with urbanization as a business model, and the role of the entrepreneurial state in materializing its aspirations.

We approach this challenge through three propositions. First, that fast cities should be examined as new 'utopias by design'. Indeed, in their elaborate claims they could also be understood as 'urban fantasies' (Watson 2014, 15) propagating 'the hope that these new cities and developments will be "self-contained" and able to insulate themselves from the "disorder" and "chaos" of the existing cities'. Second, we argue that fast cities highlight the emergence of 'entrepreneurial states' (Mazzucato 2013) that despite the current arguments of neoliberal urbanism are taking an increased interest and playing an important role in shaping mega-urbanization. The entrepreneurial state is preoccupied with 'lawfare' – the increased use of 'brute power in a wash of legitimacy, ethics, and propriety' (Comaroff and Comaroff 2006, 31) to build new cities. They use the 'metaphysics of disorder' to internalize the logics of capital and extend the executive and legislative power of the state over new territories and populations. Further, the entrepreneurial state primarily uses the visual and representational power of imagery and rhetoric to spread its message and mobilize mass aspirations around new cities. Finally, we propose that these fast cities built by entrepreneurial states actually materialize at the scale of regional urbanization rather than as metropolitan growth. They lead a new phase of mega-urbanization across the global south that while embedded in a modernist developmental legacy of urban planning, also scales up to economic and industrial corridors. In doing so, new cities bypass the pressing challenges of existing megacities to create new townships and hence risk becoming premium urban enclaves. Taken together, these three propositions highlight the contradictions between intended and unintended outcomes of fast cities and point to their fault lines between state sovereignty, capital accumulation and citizenship.

We elaborate on these arguments below.

Fast cities as 'new' urban utopias

The utopian impulse at the heart of so many experiments in city-building has always proved disappointing, if not downright disastrous, in the actual flesh and stone. Much has been written about why this is so – perhaps enough to discourage any further attempts at utopian thinking about the city. But the utopian impulse is, and will hopefully remain, an irrepressible part of the human spirit.

(Sandercock 2003, 2)

If fast cities are projected as a solution to the current urban crises, then they are in their very conceptualization a form of urban utopia by design. This is not necessarily a new claim, since the history of urban planning is a history of urban utopias. Examples of urban utopias and debates around their 'failure' abound in the context of modernist urban planning in the west (Fishman 1982). Modernist urban utopias attempted in some way or another to deal with the contingencies of our times – nuclear fallout, resource depletion, ecological degradation and so on – through technological modernism. Buckminster Fuller's concept of *Spaceship Earth*, Archigram's concept of the technologically advanced *Plug-in city*, Frank Lloyd Wright's *Broadacre City*, Le Corbusier's *Radiant City*, Paolo Soleri's ecological architecture of *Arcosanti*, Richard Register's *Eco-city* and in more recent years Bill Dunster's *Zero Energy Development* all constitute ways of radical rethinking through design, technology and urban planning towards a new city of the future. Critics of these models of city-making highlight the anthropocentric and androcentric focus of utopian city-making projects, based on their largely idealistic and utopian imaginations. Others note that these utopian visions of the city produce new kinds of social exclusions across race, class, gender, ethnicity and so on.

In the global south, too, utopian modernist planning is often positioned against colonial urban development. Colonial cities that used the trope of modernity to create 'white towns' in Kolkata, Pondicherry, Kinshasa, Johannesburg and so on were looking to create exclusionary social utopias around racial segregation. In postcolonial contexts, however, the trope of 'modern cities' was reinforced to make a break from colonial planning and its associated social injustices. This was evident in the building of several new cities – Chandigarh (Kalia 1990) and Brasilia (Holston 1989) being the most commonly cited examples. These cities were large-scale public-funded projects that seemingly began from a tabula rasa to create a city where rationality rather than tradition represented its urban design, where equality rather than difference shaped social relations, and most importantly, where a new kind of 'equal' society could be created through a rationalization and universality of the city's physical fabric. Yet contrary to popular belief, Chandigarh was not planted on 'empty' harsh plains waiting to be inhabited. For the first phase of Chandigarh, 8,500 acres of fertile land, consisting of 17 villages, were acquired in one go under the Land Acquisition Act of 1894. In another few years 24 additional villages dotted with agricultural land and mango groves were acquired. Much of the conflict that surrounded this land acquisition and the displacement of farmers (landed and not-landed) does not form part of the dominant narratives surrounding the history of this planned city (Sabhlok 2016). Chandigarh and Brasilia and other such postcolonial cities are now seen as 'blueprint utopias' (Holston 1989) since the production of new social inequalities rather than the erasure of historic social differences became one of their most persistent and unintended consequences (Shaw 2009).

Fast cities are new forms of utopias that are extensions of these earlier planning models. They combine speed, technology and planning as a utopian response to current crises. Speed is embodied not just in their conceptualization, but also in their governance, growth models and the production of entrepreneurial citizens. Like other forms of cultural production, fast cities too imagine different urban futures (often involving the use of technology) that can be set up at 'push-button' speed. This explicit coupling of urban planning and technology as a 'high modern ideal' (Graham 2000) has been around since the 20th century in producing the ubiquitously 'networked city' (Castells 1996), which was seen to speed up connectivity, surveillance and governance. These cities embodied utopian ideals of technology-driven efficiency – a utopia that was more a rhetorical and ideological device than a practical reality. Variants of this new urban form intrinsically linked to and produced from technology have been called *wired city* (Dutton et al., 1987), *telecity* (Fathy 1991), *e-topia* (Mitchell 1999), *aerotropolis* (Kasarda 2000), *intelligent city* (Komninos 2002), *sentient city* (Shepard 2011) and *smart city* (Marvin et al. 2015). Physically these variants are an exaggeration of the sprawl and suburbia that characterize urban planning of the 20th century, but they are also cities that are largely detached from their local and regional contexts, despite claims to the contrary.

Here we need to clarify the distinction between state utopianism or 'utopias by design' and the 'experimental utopias' of Lefebvre. There is wide agreement that state-led utopian thinking as referred to in the quote by Sandercock above has produced disastrous results in modernist architecture and planning. As Pinder (2015) notes, this is different from the revival of utopian thinking in 'creative praxis' which has its roots in revolutionary thinking, political struggles and grassroots imaginations. The fast cities we refer to in this book are largely state and corporate-led totalitarian visions which are undergoing a global revival under the dominant rhetorics of 'crises'. Their utopian conceptualization promotes an accelerated process of innovation, entrepreneurialism and economic growth, rather than experimentation with alternative imaginings to the dominant logics of capital.

The first three chapters in this book articulate this analytical concern with speed, temporality and utopian urbanization in different ways. Examining the logic of 'instant urbanism' in the African continent in Chapter 2, Murray notes that while Africa has so far been left out of planned urbanization, a new kind of 'rogue urbanism' (Pieterse and Simone 2013) is now defining its political and social landscape. This instant urbanism defines the 'dominant political and ideological practices of power regulated through their global connections'. Across Africa, the rise of a number of 'parallel cities', 'urban islands', 'gateways' and 'infrastructure cities' opens up new opportunities for innovative design and architectural experimentation. Murray orders these into broad typologies of a) masterplanned multifunctional satellite cities, b) Transportation and logistics hubs, c) Themed entertainment destinations, d) Themed retail destinations, and e) Hi-tech cities. He concludes that the diversity of these fast cities is in effect a 'reordering of urban temporality' by actively engaging in 'innovative strategies designed to bypass broken-down infrastructure, overcrowded thoroughfares and unworkable regulatory regimes'.

Following on from Murray in Chapter 3, Watson calls these new African cities 'urban fantasies'. She notes that not only are African cities urbanizing later, they are also doing so at a much faster pace than other continents. Watson argues that the accelerated pace of urbanization has also accelerated the exacerbation of capital accumulation and associated social inequalities. Thus in urban Africa, speed comes at the cost of social justice. She concludes that instead of learning from past mistakes of colonial governance, these new cities emulate colonial grand utopian dreams of modernity and development while simultaneously rejecting their colonial links. Her predictions about Africa's future spell dark dystopian juxtapositions of living in poverty alongside the 'Dubaisation' (Choplin and Franck 2010) of new city projects.

'Speed kills' is the conclusion Cugurullo draws in Chapter 4. Exploring 'fast regulation' in the Emirates, he argues that the desire to speed up the making of the new eco-city of Masdar in the UAE made the 'core principles of sustainability' subservient to the need to generate new business investment in the region. Thus speed came at a price and the price has been ecological devastation in the region. Arguing that sustainability and business are not always complementary interests, Cugurullo claims that although conceived as an eco-city to respond to the crisis of urbanization and climate change, the Masdar project brought about irreversible depletion of environmental resources in the region. For Cugurullo, speed has killed accountability in Masdar since its claims have not been supported by post-project monitoring. Taken together, Cugurullo, Murray and Watson outline how the speed of new city-building across Africa and the Gulf transfers risk and precarity associated with speed to social and environmental landscapes. They highlight the tenacious but often hidden links between speed and capital accumulation in the construction of new cities. Their chapters thus highlight what is lost in the fetishization of speed and what is at stake when speed becomes the priority for urbanization.

'Entrepreneurial' states and 'slow' democracy

In a recent speech, the Indian Prime Minister Narendra Modi announced that India will lay down a 'red carpet not red tape' for foreign investors (PTI 2015). This was related to his announcement of making business free of bureaucracy for foreign investors by introducing Foreign Direct Investment (FDI) in building and construction, speeding up environmental clearances, and fast tracking infrastructure development projects. This sector is key to the Indian economy as mega-urbanization requires the constant building of new cities and infrastructure. Chinese cities have also announced an 'open-door' policy to lure in investors and pay for new city-building. Global capital forces in the form of FDI permeating into the domestic political landscapes shows how the production, consumption and consequences are deeply entrenched in the priorities of the market. Several other states in the global south are now pursuing similar policies to actively shape and transform the relations between capital and production at the urban and regional scale.

What we are observing now is a key transformation in urbanization strategies in the global south as sovereign states and not only cities (as previously understood) are becoming more entrepreneurial and creative in their strategies of accumulation. As Pieterse and Simone (2013) argue in the context of Africa, current urbanization indicates a set of evolving power dynamics between an entrepreneurial state and a range of other actors that take us beyond the conventional fixation on neoliberalism. This is in a context where the postcolonial state is facing unprecedented challenges in terms of its sovereignty and legitimacy – both within its national territory as well as in a global arena – globalization, climate change, migration, rapid urbanization, grassroots activism and so on. In responding to these challenges, the state has recast its earlier goals and aspirations of modernity and development into an ideology of 'entrepreneurial urbanization' (Datta 2015a) – a mould that Hsing (2012) calls 'urbanization of the state'.

Neoliberalization is only a partial description of the diverse and varied processes through which new cities are emerging. There is now a clear emergence of entrepreneurial states in the global south which are 'market players and can even use market instruments to achieve hidden political agendas' (Xu and Yeh 2005, 284). These entrepreneurial states show some important convergences with and divergences from the notion of the entrepreneurial city (Harvey 1989). While we understand 'entrepreneurial cities' (Harvey 1989, Jessop and Sum 2000) as those which pursue private capital at all costs via speculative experiments in urbanism, we also note that western versions of entrepreneurial cities do not completely capture the complexities and diversities of entrepreneurial urbanism in the global south. For example, in cities such as Manchester and Leeds (Haughton and Williams 1996), the long-established strategy of urban renewal indicated a shift from managerialism to entrepreneurialism (Harvey 1989) that often relied on the active production of a creative class (Peck 2005). But western cities essentially see city-centric strategies as a reflection of the wider shift from a welfarist to neoliberal state, which cannot be used as a 'template' for postcolonial urbanism (Ghertner 2014). As Crossa (2009) notes, entrepreneurial cities in the south are often framed around the exclusion of informality, which dilutes several key linkages of power between the state and other stakeholders. Wu (2003, 1694) correctly points out in the case of China that 'marketisation in the post-socialist transition has brought various entrepreneurial activities within the city. But this is not equivalent to the "entrepreneurial city"'. In the Gulf region, too, Acuto (2010, 272) argues that a 'centralized and hyper-entrepreneurial approach' from the Emirati state has characterized Dubai's attempt to establish 'symbolic power' and remake itself as 'the image of the 21st century metropolis'. We argue that these radical developmental shifts in the postcolonial state have produced what Roy (2011) calls 'homegrown neoliberalism' that is geared towards urbanization. These processes indicate that urban entrepreneurialism in the global south is often mediated through the entrepreneurial state or entrepreneurial region, rather than competitive entrepreneurial city strategies alone.

This does not imply that state entrepreneurialism is completely new in the global south, but rather that the production of fast cities is now a much more state-centric entrepreneurial intervention than it ever was before. As Wu argues (2003, 1678), 'intervention is achieved through a huge state bureaucratic system and the degree of the state's involvement in directing production processes is pervasive'. Not just in China, but in several countries of the global south, the state is no longer a neoliberalizing bureaucratic establishment withdrawing welfare and fixing market failures; rather, is an active player in producing 'innovation' and enticing private capital into its new cities. The entrepreneurial state (like the entrepreneurial city) is marked by its strategies, discourses and images (Jessop and Sum 2000) to become a 'market player', but is also engaged in an ideology of urban entrepreneurialism that seeks to reinforce and legitimize sovereign power.

We find then that the notion of 'entrepreneurial state' (Mazzuccato 2012) more appropriately captures the ongoing urban transformations in the global south that is producing fast cities. This stems from our observation that the explosion in Indian, Chinese and Middle-Eastern urbanization points to a form of state entrepreneurialism that has so far remained largely under-researched in urban studies. This observation is wide ranging across democratic, post-socialist, autocratic and monarchic states which have all made different innovations in their development policies to diversify their economy and spur economic growth via city-building. However, while Mazzucato frames the entrepreneurial state in largely progressive terms, the chapters in this book suggest that the shift towards state entrepreneurialism in the south, like entrepreneurial cities, often (re)produces and reinforces social and spatial inequalities through their active pursuit of capital (Harvey 1989, Peck 2002). This is largely because the differences between the state space and the spaces of the private sector are becoming more and more indistinct in much of the global south. As Ponzini (2011, 254) argues in the context of the Middle-East, 'the separation between public and private sectors . . . is practically non-existent because the actors have key positions in public decision making and in the management of private companies'.

In countries like India and China, too, the political elite often hold shares in major construction and infrastructure companies, and the private sector are often funders of major political campaigns for leadership. Thus the building of Masdar and other new cities in oil-rich economies is possible because 'petrodollars' can fund mega-urbanization within a democratic vacuum (Moser et al. 2015). The building of Songdo or Dongtan similarly highlights the desire of a post-socialist state to diversify its industrial economy, which can bypass several democratic processes to dispossess existing residents from their land. The construction of Quito International Airport in Equador highlights a state-supported entrepreneurial approach to urban infrastructure projects (Carrion 2015). The building of 100 smart cities in India can similarly be made possible by making radical changes to urban development policies and by marketing the idea of a 'Digital India' to its urban youth, hungry for increased economic prosperity (Datta 2015b). State ideologies and priorities drive the entrepreneurialism and innovation in these city-building projects, which reinforce inequalities at the local and regional scales.

In these contexts, 'development' as both a logic and solution to the crisis of urbanization becomes a tool for asserting the material and symbolic power of the state over its citizens. In earlier postcolonial contexts, the state was expected to be an engine of development, a harbinger of industrialization, building new towns, physical infrastructure, and social and cultural institutions that aimed to remove poverty. State leaders such as Nehru imaginatively applied the notion of state developmentalism through the building of new cities like Chandigarh that staked their postcolonial claims to modernity. Development enacted by the contemporary entrepreneurial state is an extension of this remaking of modernity. It uses city-building and urban development as the primary site of its political legitimacy and sovereign rule as a 'regime of accumulation' (Levien 2013). This is evident across Asia, the Middle-East and Latin America where ruling political parties assert their dominance on the back of their urban growth agendas. As Xu and Yeh (2005) suggest, this encourages the emergence of 'growth coalitions' between the state and private sector based on money-generating sectors such as construction, infrastructure and real estate. The state not only 'de-risks the private sector, but envisions the risk space and operates boldly and effectively within it to make things happen' (Mazzucato 2012, 3). It then no longer directly delivers on development projects. Rather, as we have already suggested, it gets involved in the process of manipulating territory via land readjustments and speculation.

These entrepreneurial states embody different conceptualizations of speed and inertia as evident from the next three chapters of this book. In Chapter 5, Shin examines the case of South Korea to argue that the making of Songdo city is envisioned by the state via a shift to entrepreneurialism while maintaining a strong reliance on public–private partnerships for urban development. For Shin, this cannot be compared to the entrepreneurial urbanism of the global north; rather, it needs to be seen as the territorial manifestation of a developmental state. Shin predicts that Songdo promotes a 'real-estate utopia' through state investment in the built environment, and will ultimately turn into a segregated exclusive enclave for the rich and powerful.

Continuing on the theme of state-driven urbanization, Rizzo shows in Chapter 6 that Qatar, like other Gulf countries, is making a radical shift away from its petroleum-based economy to a knowledge-based economy at a rapid pace. As he notes, it was indeed the Gulf States which prompted Bagaeen (2007, 174) to use the term 'instant urbanism' to differentiate them through speed, time and duration from western urbanism. Examining Qatar Foundation's Education City, Rizzo finds that 'knowledge megaprojects' in the Gulf are now the key sites of state investment that drive its rising geopolitical ambitions. This is despite the fact that the political and financial commitments of the state have to be higher in these projects to compensate for the potential loss of consumption-related economies in the knowledge city. Rizzo notes that the shift away from petro-urbanism and petro-politics to 'cleaner' urbanization is actually fuelled by a 'reverse colonization' of western assets through news media, membership in international organizations and knowledge partnerships. The Qatari state relies on high-speed connectivity, fast communication and fast decisions across knowledge partners in the west to create and transition to seemingly 'post-carbon' economies. To that end, knowledge megaprojects in the Gulf are stirred, sponsored and implemented by a majority, government-linked shareholder. Yet in the end, these knowledge cities 'replicate the same pitfalls of other new cities in producing "splintered urbanism"' (Graham and Marvin 2001) across the region.

While constitutional monarchies such as Qatar or post-socialist states such as China are able to move forward relatively quickly on megaurbanization projects, constitutional democracies such as India are often seen to be 'slowed' down by statutory processes of deliberative planning and democracy. The diversity of 'speed' in the manifestation of fast cities in these different contexts can be said to arise from the adoption of different forms of entrepreneurialism by the state and regional institutions.

In Chapter 7, Kundu examines the managerial role of the local/regional state in the transformation of Kolkata's urban peripheries for building Rajarhat new town. Kundu's conceptualization of the entrepreneurial state is critically at the scale of the region rather than the nation. The regional state of West Bengal in India with relative autonomy from the federal state induced massive land transformations along the urban peripheries through a logic of 'territorial flexibility' put to use by creating legal ambiguities around land use, acquisition and transfer. As Kundu suggests, the state's zeal for accelerating construction of the township was stalled substantially through 'blockades' (Roy 2011), which underlined the duality of speed and inertia through grassroots resistance to the 'intrusion of the city into the village'. Kundu concludes that the 'ways in which groups that face removal or displacement by strategies of entrepreneurial governance negotiate, resist or even subvert' the rhetorics and practices of speed highlight the fault lines along which fast cities are conceived. Rajarhat can thus be understood as 'centre stage in the politics of accumulation and dispossession today' (Hsing 2012), the flip side of fast cities that have made 'peasants the final frontier in city-making' (Goldman 2011).

These three chapters of the book then note how despite the wide variety of state structures across the global south, and despite their radical trans-formations in a neoliberal era, the state continues to broker a strong role in regional urbanization. Even when it appears starkly absent, the state maintains an active role in politics, representation and rhetorics that direct public discourses, as well as through its more concrete institutions, laws, policies, planning and bureaucratic setups. Taken together, they suggest how city-building is now the new mechanism of state-building, modernity and globalization in the postcolony.

Mega-urbanization and masterplanning

Every technical practice is a social practice, every technical practice is soaked in social determination. But this does not present itself as such: it claims autonomy, innocence, a technical rationality founded on science. This rationality subtends the ideology of faith, which imposes itself on our society as morality, wherein technical practices, separated from social reason, become a technique of the social, and more precisely of social manipulation, and therefore a technics of power.

(Baudrillard 2006, 51)

For Baudrillard, technical practice is a method of differentiation and distinc-tion between and across 'zones of privilege'. The power of technical practice lies in its simultaneous practice of mythmaking to obscure the subjectivity and present it as if it was a rational outcome, a science. Technical practice for Baudrillard lies in perpetuating social techniques of segregation, exclusion and discrimination through the perceived democratization and rationalization of a technique of the social. Entrepreneurial states engage directly in the technical practice of planning fast cities.

If technical practice is a terrain of power, then masterplanning is its currency. The masterplan has always been a blueprint for urbanism, deployed as a technique of governance. It presents both its democratic and rational aspects which hide the subjective dimension of politics and culture. As a tool of urban development, the masterplan is produced and legitimized by a group of experts who have been variously called 'agents of the state' (Peck 2002) or 'Global Intelligence Corps' (Olds 2001). By this we understand the entire range of planners, policy makers, consultants, international design firms, energy certifiers, IT consultants and others with standard assessment tools, glossy brochures and impressive presentations who 'sell' and legitimize the idea of the masterplan to those who are already constitutive of the networks of privilege and power – the political and social elites. The masterplan rationalizes the mega-urbanization dreams of the state to its citizens as if it was the only rational solution to urban crises, and in doing so it mobilizes mass dreams and future aspirations. The elaborate mythmaking around the urgency of the masterplan feeds off the circulation of elaborate 'rational' predictions around migration, urbanization and climate change. They present a moral imperative to act upon the crises, to then perpetuate as an ideology of mega-urbanization and masterplanning.

Fast cities manifest at a very distinct moment of governmentality of the masterplan. While fast cities can be considered as utopian solutions to the 'modernist watchword' (Martin 2010, xi) of urban crisis, they are in reality key to the strategies of entrepreneurial governance of the state. They initiate executive action through the legal, extra-legal and brute power embodied in the masterplan. This new mode of governmentality highlights a shift from the state's earlier Foucauldian fixation with controlling populations to a renewed interest in the manipulation of territory (Schindler 2015). Schindler (2015) notes that this is due to a fundamental disjunct between capital and labour – the former invested in real estate and the latter struggling to make a living wage. This disjunct is most pronounced when this involves the creation of new cities since it involves the reach of the postcolonial state beyond what was possible before – in the seizure and control of new territories outside the 'urban'. The masterplan can be seen as a close ally of the state in this new 'governance of territory' (Schindler 2015, 7). It provides a blueprint for governing through a) measurement techniques – satellite imagery, land-use surveys, topographical surveys, floodplain maps, revenue and taxation records; b) impact assessments that predict and minimize the depletion of natural and social resources; and c) design and zoning of different land uses for commercial, office, retail and residential spaces in new cities. In other words, it is not just in the production of new 'urban' territories but rather in the fundamental transformations in the meaning of the 'urban' that the masterplans of fast cities become a tool of governmentality.

The rhetoric and imageries of speed, time and scale constitute the legitimization of fast planning. Thus masterplanning is as much about the visual power of representation as it is about the techne of control. Speed, crisis, urgency and growth are regularly translated into aspirations for the state and its citizens through glossy brochures, flythrough videos, animated walkthroughs, colourful pie charts and graphs. Its citizens claim to differentiate themselves from their colonial counterparts through the emergence of new kinds of postcolonial subjectivities around technology-driven participatory spaces, yet they present the same kind of assumptions and fantasies around disembodiment and control 'that have marked science, technology and mass communications in the west' (Grosz 2001, 41). The language and visual imagery coalesce to become a form of 'corporate storytelling' (Soderstorm 2014) of the merits of the masterplan that circulate as solutions to a global urban crisis. Their claims to sustainability, growth and development perpetuate simulated experiences of urban life around prosperity, lifestyle and innovation that are unrealizable in the long term.

We argue then that masterplanning and mega-urbanization follow two distinct techniques of governmentality. First, a key attribute is the ability to intervene and make wide-ranging territorial manipulations at a regional scale. These manipulations need the accelerated establishment of new laws, planning mechanisms, regulations, bureaucratic processes and implementation of new city and mega-infrastructure projects, which 'seems to reflect the everyday roil of incrementalist fast-policy adjustments, managed within tight fiscal and ideological parameters, and the rescheduling and displacement of crisis tendencies' (Peck 2002, 400). While extra-legal practices are an intrinsic technique of the state in implementing the masterplan, a new mode of governmentality fixated on 'lawfare' (Comaroff and Comaroff 2006) has emerged in recent years. This 'lawfare' materializes the masterplan through a renewed focus on planning (Nair 2015). As Nair notes in the case of India, while the rule of law has traditionally been seen to delay and defer large infrastructure projects, the rule of law is now used precisely to maintain a consensus around fast cities as engines of growth. This consensus earlier included the state and private sector, but now includes innovation in the making and interpretation of law by the judiciary and legislature. Nair argues that by recasting land speculation as in the 'public interest', the state makes particular innovations in masterplanning. This has reinforced the role of the masterplan in reclassifying land as real estate, transforming citizenship rights and severing the connections between land and territory for vast swathes of the rural population.

Second, we argue that the masterplans of new cities are part of an elaborate representational work of constructing the commons as 'terra nullis' (Jazeel 2015). Land fictions in the global south now constitute the most 'elementary

19

extractions' (Sassen 2014) of a regional urbanization embodied in fast cities. The transformation of land as commons or land as livelihoods into land as commodity obscures the contested local struggles around dispossession and large-scale land grabs. In India, for example, land has been a key space of inertia in the speed and scale of its mega-urbanization and fast city aspirations (Datta 2015a). Fierce debates have been raging across political and civil society around amendments to a colonial Land Acquisition Law that requires consent from those whose land is to be acquired for 'public purpose'. The proposed amendment to this law, which plans to do away with the consent clause and replace this with market rate compensation, brings to question precisely the issue of what constitutes 'public' property and how that might be legislated. In doing so, masterplanning and mega-urbanization in fast cities have not only exposed the contentious politics between private property and public goods, but also reinforced the inequalities between urban and rural citizenships.

The three chapters in this final section of the book by Hudalah and Firman, Lane and Morera suggest how a model of 'masterplanning for business' underlines speed as a dominant mode of contemporary governmentality. All three chapters argue that mega-urbanization and masterplanning are justified by the entrepreneurial state through a duality of crises and growth. Hudalah and Firman argue in Chapter 8 that new city-building around Jakarta is spurred by industrial suburbanization along its peripheries. This suburbanization is induced by the workings of FDI that seeks to boost the economy of mega-urban regions in South-East Asia. This has led to the emergence of a smattering of industrial parks, Special Economic Zones and new townships. In essence, these are scaled-up versions of earlier gated communities which in the absence of large-scale infrastructure investments have become fragmented and isolated from each other. But they also induce industrial suburbanization around Jakarta that pays less attention to resolving the 'problems' of the mega-city than reorganizing land around the city in the service of real estate. Mega-urbanization around Jakarta is thus also a process of mega-suburbanization where the city itself is now a mere reference point for the manipulation of territorial power in the region.

In Chapter 9, Lane similarly examines the power–knowledge networks in Lusaka's new vision 2030 masterplan which effectively aims to urbanize Lusaka's surrounding regions through a series of satellite cities. Similar to Herbert and Murray's (2015) findings in the case of Johannesburg's satellite cities, Lane too finds that although Lusaka's new masterplan sought to rebuild the existing city in the image of a 'garden city', its expansionist ideology in the region was brought in through the backdoor of the masterplan. In Lane's argument, the 'new' city is represented through the mega-urbanization of Lusaka's surrounding region, using Lusaka merely as a reference point. Legitimized as the 'answer to Lusaka's sustainability concerns', the masterplan rationalized local anxieties over urban informality and economic decline in Lusaka's peripheries by large-scale land-use transformations and infrastructure expansions along development corridors reaching deep into its hinterlands. Although explicitly claiming to address the concerns of the existing city, the masterplan is a case in point of a 'bypass urbanization' (Bhattacharya and Sanyal 2011) where the 'promotion of a dual core CBD moves economic activity away from the old central areas and clusters it in new regions with "greater potential" – for example along the main road leading to the international airport'. The production of the masterplan is legitimized via the binary construction of fast and slow, which overlapped with the rationalization of the masterplan and institutional incompetence respectively.

Following on from the themes of masterplanning for sustainability, Morera examines the visual power of planning through the political imagery of two Chinese eco-city projects – Dongtan and Tianjin. By highlighting the different paths taken by these two eco-city projects, Morera argues that masterplanning is now the key site of negotiations of the contradictory claims made by these projects. But, in his study, masterplanning is not just planning practice; it also works as political propaganda around 'ecological urbanization' of the state. This propaganda is perpetuated through carefully selected images – colours, figures and compositions in urban landscapes that comprise politically charged messages of representation. Morera therefore concludes that the visualizations of masterplans as local and global forms of communication have worked as political tools of the state, 'with or without the agreement of planners'.

Taken together, the three chapters in this section underline the techne of governmentality of the entrepreneurial state through the manipulation of territory represented in the masterplan. This manipulation might refer to the material shifting of land from private ownership to state-led acquisition for 'public interest' or the representational power of the masterplan in shifting public opinion through rhetorics and imagery. Overall they highlight the flipside of speed and the relative construction of 'fast' and 'slow' embodied in the masterplan's quest for a business model of urbanization.

Towards repetition and difference

Fast cities defy the systems and rationalities of postcolonial urban studies and therefore pose a challenge to their legitimacy as an analytical category. They focus on speed as a strategy of innovating, competing and 'leapfrogging' into 'sustainable urban futures'. They do this through the active construction of temporal binaries between fast policy and slow government, free markets and state bureaucracy, masterplanning and local consultation, state vision and local democracy. But speed itself is relative and the diversity of fast cities in different contexts suggests that they cannot be cast neatly within 'concentrated and extended' (Brenner and Schmid 2011) modes of urbanization or indeed within the lens of inter-referencing or 'worlding' (Roy and Ong 2011). The parameters of their making are temporal because they need to suit the imperatives of current crises. Their legitimacy derives from a rhetoric of urgency which calls for a speeding up of the processes of bureaucracy, planning and democracy which are traditionally seen to slow down urbanization. Fast cities claim to deal with the present by seizing the future. And the future is the space of scenarios, projections and utopian claims. The future cannot be measured and called to account since it has not yet materialized. Analysing fast cities then calls for a new mode of doing urban studies that needs to take account of their duality of speed and inertia as a critical aspect of their manifestation.

This book therefore seeks to capture the rhythms of a 'moving but determinate complexity' (Lefebvre 2004, 12), of a postcolonial return to built form as a solution to crises. Fast cities reference the quantitative aspect of time that defines our urban crises – stock market crashes, migration trends, climate data modelling, probability scenarios and predictions, among others. But this book argues that they also bring together the qualitative aspects of time – rhetorical techniques, knowledge coalitions, middle-class desires, postcolonial anxieties and the techne of sovereign power concealed within the ideology of urbanization. The rhythms of fast cities capture the rational and subjective aspects of time, speed and duration that bring together the cyclical and linear aspects of postcolonial urban planning. These were captured by Lefebvre (2004, 9) through a juxtaposition of his 'methodologically utilisable categorisations' as below.

> Repetition and difference
> Mechanical and organic
> Discovery and creation
> Cyclical and linear
> Continuous and discontinuous
> Quantitative and qualitative . . .

The chapters that follow in this book highlight that we need to add further methodological categorizations to this list. They are purposefully dialogical to show the cyclical rhythms of speed and inertia embodied in the construction of new cities. They are heuristic tools through which we approach the question of what is 'new' in fast cities, and why we should pay attention to them.

> Utopias and dystopias
> Iconic and prosaic
> Masterplans and tactics
> Capital and expulsions

Governance and citizenship
Privilege and exclusions
Law and injustice
Imagined cities and lived urbanism

The *Fast cities* in the rest of the book show us how cities built on sand are now the ideology of urban futures in the global south. They show us how grandiose ideas of utopian cities emerge, circulate and are scaled up in particular contexts and therefore what is at stake for the state and its citizens who aspire to them. *Fast cities* in this book show us how the 'rationalities' of crises and growth obscure from view their embedded fault lines in citizenship, identity and belonging. *Fast cities* show us how making and not living in these new cities is the techne of postcolonial rule in current times.

This book then concludes with a vision and provocations for 'slow' cities and decelerated urbanism. We argue that speed itself cannot be a driver for innovation and growth. Nor can it justify claims to counter the inertia in bureaucracy, planning and democratic planning processes. Rather, by reflecting on the chapters in this book, we argue that a notion of 'slow cities' should prioritize reflexive planning and urbanization in the global south. Rather than a binary construction of fast and slow cities, we argue for a prioritization of time and duration in the projections of future urbanism. The provocations calls for a) slow urbanism, b) slow governance, c) democratizing the commons and d) alternative utopias. Taken together, we end this book by staking claims to a different kind of future that places the citizen at the centre stage of postcolonial urbanism.

References

Acuto, M. 2010. "High-rise Dubai urban entrepreneurialism and the technology of symbolic power." *CITY* 27: 272–284.

Atkinson, R. and Blandy, S. 2005. "Introduction: international perspectives on the new enclavism and the rise of gated communities." *Housing Studies* 20.2: 177–186.

Baudrillard, J. 2006. *Utopia Deferred: Writings for Utopia* (1967–1978). Cambridge, MA: MIT Press.

Bhattacharya and Sanyal, B. 2011. "Bypassing the squalor: new towns, immaterial labour and exclusion in post-colonial urbanisation." *Economic and Political Weekly*, July 30, 41–48.

Brenner, N. and Schmid, C. 2011. "Planetary urbanisation," in Mathew Gandy (ed) *Urban Constellations*. Berlin: JOVIS Verlag, 10–13.

Breitung, W. 2012. "Enclave urbanism in China: attitudes towards gated communities in Guangzhou." *Urban Geography* 33.2: 278–294.

Bunnell, T. and Das, D. 2010. "Urban pulse – a geography of serial seduction: urban policy transfer from Kuala Lumpur to Hyderabad." *Urban Geography* 31(3): 277–284.

Caprotti, F., Springer, C. and Harmer, N. 2015. "Eco for whom? Envisioning eco-urbanism in the Sino-Singapore Tianjin Eco-city, China." *International Journal of Urban and Regional Research* 39(3): 495–517.

Carrion, A. 2015. "Megaprojects and the restructuring of urban governance: the case of the New Quito International Airport." *Latin American Perspectives*, DOI: 10.1177/0094582X15579900

Castells, M. 1996. *The Rise of Network Society: The Information Age: Economy, Society, and Culture* (2010 ed.). Chichester: Wiley-Blackwell.

Chakrabarty, D. 2000. *Provincializing Europe – Postcolonial Thought and Historical Difference*. Princeton: Princeton University Press.

Chen, X., Wang, L. and Kundu, R. 2009. "Localizing the production of global cities: a comparison of new town developments around Shanghai and Kolkata." *City and Community* 8(4): 433.

Choe, K., Laquian, A. and Kim, L. 2008. "Urban development experience and visions: India and the People's Republic of China." ADB Urban Development Series, Manila, Asian Development Bank. Accessed on 13 October 2013 from http://citiesalliance.org/sites/citiesalliance.org/files/ADB_Urban-Visions.pdf

Choplin, A. and Franck, A. 2010. "A glimpse of Dubai in Khartoum and Nouakchott: prestige urban projects on the margins of the Arab world." *Built Environment* 36(2): 192–205.

Comaroff, J. and Comaroff, J. 2006. *Law and Disorder in the Postcolony*. Chicago and London: University of Chicago Press.

Crossa, V. 2009. "Resisting the entrepreneurial city: street vendors' struggle in Mexico City's historic center." *International Journal of Urban and Regional Research* 33.1: 43–63.

Datta, A. 2012. "India's Eco-city? Urbanisation, environment and mobility in the making of Lavasa." *Environment and Planning C* 30(6): 982–996.

—— 2015a. "New urban utopias of postcolonial India: entrepreneurial urbanization in Dholera smart city, Gujarat." *Dialogues in Human Geography* 5(1): 3–22.

—— 2015b. "100 smart cities, 100 utopias." *Dialogues in Human Geography* 5(1): 49–53.

Dutton, W., Blumler, J. and Kraemer, K. (eds.) 1987. *Wired Cities: Shaping the Future of Communications*. New York: G.K. Hall.

Fathy, T. 1991. *Telecity: Information Technology and its Impact on City Form*. New York: Greenwood Press.

Fen, W., Quilun, C. and Bing, H. 2016. "De/constructing urbanization," in Meiqin Wang (ed) *Urbanization and Contemporary Chinese Art*. London: Routledge.

Fishman, R. 1982. *Urban Utopias in the Twentieth Century: Ebenezer Howard, Frank Lloyd Wright, and Le Corbusier*. Cambridge, MA: MIT Press.

Garreau, J. 1992. *Edge City: Life on the New Frontier*. Grantham, NH: Anchor Books.

Ghertner, A. 2014. "India's urban revolution: geographies of displacement beyond gentrification." *Environment and Planning A* 46: 1554–1571.

Glasze, G., Webster, C. and Frantz, K. (eds.) 2006. *Private Cities: Global and Local Perspectives*. New York: Routledge.

Goldman, M. 2011. "Speculative urbanism and the making of the next world city." *International Journal of Urban and Regional Research* 35(3): 555–581.

Graham, S. 2000. "Constructing premium networked spaces: reflections on infrastructure network and contemporary urban development." *International Journal for Urban and Regional Research* 24(1): 183–200.

Graham, S. and Marvin, S. 2001. *Splintering Urbanism: Networked Infrastructure, Technological Mobilities and the Urban Condition*. Routledge: London.

Grosz, E. 2001. *Architecture from the Outside: Essays on Virtual and Real Space*. Cambridge, MA: MIT Press.

Harvey, D. 1989. "From managerialism to entrepreneurialism: the transformation in urban governance in late capitalism." *Geografiska Annaler. Series B. Human Geography* 71(1): 3–17.

Herbert, C. W. and Murray, M. J. 2015. "Building from scratch: new cities, privatized urbanism and the spatial restructuring of Johannesburg after Apartheid." *International Journal of Urban and Regional Research*, DOI:10.1111/1468-2427.12180

Hogan, T. and Houston, C. 2002. "Corporate cities: urban gateway or gated communities against the city? The case of Lippo, Jakarta," in T. Bunnell, L. B. W. Drummond and K. C. Ho (eds.) *Critical Reflections on Cities in Southeast Asia*. Singapore: Times Media Private Ltd, 243–264.

Holston, J. 1989. *The Modernist City: An Anthropological Critique of Brasilia*. Chicago: University of Chicago Press.

Hsing, Y. 2012. *The Great Urban Transformation: Politics of Land and Property in China*. Oxford: Oxford University Press.

Jazeel, T. 2015. "Utopian urbanism and representational city-ness." *Dialogues in Human Geography* 5(1): 27–30.

Jessop, B. and Sum, N-L. 2000. "An entrepreneurial city in action: Hong Kong's emerging strategies in and for (inter)urban competition." *Urban Studies* 37: 2287.

Kalia, R. 1990. *Chandigarh: The Making of an Indian City*. Oxford: Oxford University Press.

Kasarda, J. 2000. *Aerotropolis: Airport-Driven Urban Development*. ULI on the Future: Cities in the 21st Century. Washington, D.C.: Urban Land Institute.

Komninos, N. 2002. *Intelligent Cities: Innovation, Knowledge Systems and Digital Spaces*. London: Routledge.

Kundu, A. 2014. "India's sluggish urbanization and its exclusionary development," in G. McGranahan and G. Martine (eds.) *Urban Growth in Emerging Economies: Lessons from the BRICS*. London: Routledge.

Lauermann, J. 2015. "The city as developmental justification: claimsmaking on the urban through strategic planning." *Urban Geography*, DOI:10.1080/02723638. 2015.1055924

Lees, L., Shin, H. and López-Morales, E. (eds.) 2015. *Global Gentrifications: Uneven Development and Displacement*. London: Policy Press.

Lefebvre, H. 2004. *Rhythmanalysis: Space, Time and Everyday Life*. London: Continuum.

Levien, M. 2013. "Regimes of dispossession: from steel towns to special economic zones." *Development and Change* 44(2): 381–407.

Lindsay, G. 2011. "City-in-a-box: are made-from-scratch metropolises the answer to Asia's urban overpopulation?". *Slate*. Accessed on 13 July 2016 from http://www.slate.com/articles/life/departures/2011/11/brand_new_asian_cities_are_made_from_scratch_cities_the_answer_to_urban_overpopulation_.html

MacLeod, G. 2011. "Urban politics reconsidered: growth machine to post-democratic city?" *Urban Studies* 48(12): 2629–2660.

Martin, R. 2010. *Utopia's Ghost: Architecture and Postmodernism, Again.* Minneapolis, MN: University of Minnesota Press.

Mazzucato, M. 2013. *The Entrepreneurial State: Debunking Public vs. Private Sector Myths.* London: Anthem Press.

McFarlane, C. 2011. "Assemblage and critical urbanism." *City* 15(2): 204–224.

McKinsey Global Institute. 2010. "India's urban awakening: building inclusive cities, sustaining economic growth." New Delhi: McKinsey Global Institute.

Mitchell, W. J. 1999. *e-topia: Urban Life, Jim—But Not As We Know It.* Cambridge, MA: MIT Press.

MLTM (Ministry of Land, Transport and Maritime Affairs). 2005. *Enterprise Cities.* Accessed on 24 October 2011 from http://enterprisecity.moct.go.kr/eng/index.jsp

Moser, S. 2010. "Putrajaya: Malaysia's new federal administrative capital." *Cities: The International Journal of Urban Policy and Planning* 27(3): 285–297.

Moser, S., Swain, M. and Alkhabbaz, M. H. 2015. "King Abdullah Economic City: Engineering Saudi Arabia's post-oil future." *Cities* 45: 71–80.

Nair, J. 2015. "Is there an Indian urbanism?" *Economic and Political Weekly,* DOI:10.5790/hongkong/9789888139767.003.0002

Narain, V. 2009. "Growing city, shrinking hinterland: land acquisition, transition and conflict in peri-urban Gurgaon, India." *Environment and Urbanization* 21(2): 501–512.

New Cities Summit. 2015. Accessed on 13 July 2016 from http://www.newcities summit2015.org/

Olds, K. 2001. *Globalization and Urban Change: Capital, Culture, and Pacific Rim Mega-Projects.* Oxford: Oxford University Press.

Peck, J. 2002. "Political economies of scale: fast policy, interscalar relations, and neoliberal workfare." *Economic Geography* 78(3): 331–360.

—— 2005. "Struggling with the creative class." *International Journal of Urban and Regional Research* 29(4): 740–770.

Pembina Institute nd. "Fast cities executive summary." Accessed on 13 July 2016 from https://www.pembina.org/reports/fast-cities-summary.pdf

Percival, T. and Waley, P. 2012. "Articulating intra-Asian urbanism: the production of satellite cities in Phnom Penh." *Urban Studies* 49(13): 2873–2888.

Pieterse, E. and Simone, A. 2013. *Rogue Urbanism: Emergent African Cities.* Johannesburg: Jacan.

Pinder, D. 2015. "Reconstituting the possible: Lefebvre, utopia and the urban question." *International Journal of Urban And Regional Research,* DOI:10.1111/1468-2427.12083

Ponzini, D. 2011. "Large scale development projects and star architecture in the absence of democratic politics: the case of Abu Dhabi, UAE." *Cities* 28(3): 251–259.

PTI 2015. "India offers red carpet to investors, no red tape, Modi tells Japan." Accessed on 13 July 2016 from http://timesofindia.indiatimes.com/india/India-offers-red-carpet-to-investors-no-red-tape-Modi-tells-Japan/articleshow/4150 8466.cms

Rapoport, E. 2014. "Globalising sustainable urbanism: the role of international masterplanners." *Area,* DOI:10.1111/area.12079

Roy, A. and Ong, A. 2011. *Worlding Cities: Asian Experiments and the Art of Being Global*. Oxford: Wiley-Blackwell.

Roy, A. 2011. "The blockade of the world-class city: dialectical images of Indian urbanism," in A. Roy and A. Ong (eds.) *Worlding Cities: Asian Experiments and the Art of Being Global*. Oxford: Wiley-Blackwell.

Sandercock, L. 2003. *Cosmopolis II: Mongrel Cities of the 21st Century*. London: Continuum.

Sassen, S. 2014. *Expulsions: Brutality and Complexity in the Global Economy*. Cambridge, MA: Harvard University Press.

Schindler, S. 2015. "Governing the twenty-first century metropolis and transforming territory." *Territory, Politics, Governance* 3(1): 7–26.

Scott, B. 2015. "These are the top 10 cities for rapid growth." Accessed on 13 July 2016 from http://www.inc.com/bartie-scott/ss/2015-inc5000-top-10-cities-for-fast-growing-companies.html

Shatkin, G. 2007. "Global cities of the south: emerging perspectives on growth and inequality." *Cities* 24(1): 1–15.

Shaw, A. 2009. "Town planning in postcolonial India, 1947–1965: Chandigarh re-examined." *Urban Geography* 30(8): 857–878.

Shepard, M. 2011. *Sentient City: Ubiquitous Computing, Architecture, and the Future of Urban Space*. Cambridge, MA: MIT Press.

Sheppard, E., Leitner, H. and Maringanti, A. 2013. "Provincializing global urbanism: a manifesto." *Urban Geography* 34(7): 893–900.

Söderström, O., Paasche, T. and Klauser, F. 2014. "Smart cities as corporate storytelling." *City* 18(3): 307–320.

Theodore, N. and Peck, J. 2015. *Fast Policy: Experimental Statecraft at the Thresholds of Neoliberalism*. Minneapolis, MN: University of Minnesota Press.

Townsend, A. 2013. *Smart Cities: Big Data, Civic Hackers, and the Quest for a New Utopia*. New York: W.W. Norton & Co.

Watson, V. 2009. "Seeing from the south: refocusing urban planning on the globe's central urban issues." *Urban Studies* (11): 2259–2275.

—— 2014. "African urban fantasies: dreams or nightmares?" *Environment and Urbanization* 26: 215–231.

While, A., Jonas, E. G. and Gibbs, D. 2014. "The environment and the entrepreneurial city: searching for the urban 'sustainability fix' in Manchester and Leeds." *International Journal of Urban and Regional Research* 28: 549–569.

Wu, F. 2003. "The (post-) socialist entrepreneurial city as a state project: Shanghai's reglobalisation in question." *Urban Studies* 40(9): 1673–1698.

Xu, J. and Yeh, A. G. O. 2005. "City repositioning and competitiveness building in regional development: new development strategies in Guangzhou, China." *International Journal of Urban and Regional Research* 29.2: 283–308.

Part I

FAST CITIES AND 'NEW' URBAN UTOPIAS

2

FRICTIONLESS UTOPIAS FOR THE CONTEMPORARY URBAN AGE

Large-scale, master-planned redevelopment projects in urbanizing Africa

Martin J. Murray

Over the past several decades, large-scale property developers have initiated an increasing number of urban redevelopment schemes designed to reshape the existing spatial landscapes of a growing number of cities in Africa. If these ambitious approaches to city building successfully move beyond the design phase and reach the stage of actual implementation, many large metropolises in Africa will be fundamentally reconfigured to more explicitly serve the aims and interests of large-scale transnational corporate investors, local business coalitions, and affluent consumers wanting to be plugged into the global economy. What makes these novel approaches to city building different from previous attempts at urban regeneration is that they involve building entirely new cities out of whole cloth rather than incrementally rehabilitating physical landscapes already in place. Dispensing with the arduous task of refurbishing the built environment of existing cities, these large-scale real estate developers – sometimes acting on their own initiative and other times operating in partnership with public authorities – have simply started over, constructing entirely new cities that are built entirely from scratch.[1]

In the futuristic fantasy projection of city builders, these master-planned, holistically designed urban enclaves are 'city doubles', *faux* doppelgängers that are the mirror (and yet polar) opposites of existing urban landscapes in Africa. By doing away with overcrowding, chaos, and congestion, these 'parallel cities' represent everything existing metropolises in urban Africa are not. Imagined as a radical alternative to the alleged 'failed urbanism' of contemporary Africa, this *tabula rasa* approach to city building signals a significant departure from conventional thinking about rebuilding urban fabric through incremental, piecemeal interventions into existing physical landscapes in need of repair (Myers and Murray 2006, Watson 2009). These new city-building projects violate the fundamental modernist principles of erasure and re-inscription, or the creative destruction of the built environment. Instead, this novel approach to 're-urbanism' involves building entirely new cities *de novo*, without any intermediary steps. Constructing entirely new physical landscapes enables city builders to bypass and circumvent the seemingly intractable problems associated with the current state of urbanism in Africa: neglected infrastructure, overcrowded streetscapes, traffic gridlock, the threat of crime, and woefully inadequate regulatory regimes (including the virtual lack of land-use planning and weak code enforcement).

Sometimes located outside the existing boundaries of major metropolises, these new 'parallel cities' promise to deliver high-quality infrastructure, up-to-date services, and 'first-world' lifestyle options, in an aesthetically pleasing, highly regulated, carefully monitored, and safe-and-secure environment. These self-contained, mixed-use urban enclaves provide the familiar blend of programmatic ingredients, including securitized office parks, upscale shopping malls, exclusive residential accommodation, and a wide range of entertainment-leisure activities. These 'parallel cities' are connected both physically and symbolically to the leading metropolitan centres of global power, but delinked from their immediate surroundings and the city life right outside their boundaries. These emergent urban islands of high finance catering for the business class and for the excessive consumption of affluent residents signify what amounts to a spatial restructuring of literally dozens of cities across rapidly urbanizing Africa. By bringing together mixed-use facilities, walkability, and holistic planning borrowed from the principles of New Urbanism in combination with the (high)-modernist stress on easy circulation, rational order, and efficient use of space in a single location, they seem to offer the best of all possible worlds. Above all, these newly constructed 'parallel cities' aspire to become key nodal points or 'gateways' – that is, new globalizing transnational spaces – in the wholesale spatial restructuring of global metropolitan networks (Sigler 2013).

These energetic efforts at re-urbanizing cities in Africa reflect dominant political and ideological practices of power regulated through their global connections. Their true intent to become spatially located 'profit-making machines' is camouflaged in the legitimacy of the local through promises of 'job creation', the emancipatory rhetoric of sequestered lifestyle choices, and the pledge of eco-friendly sustainability. These proposed satellite cities operate under the umbrella of new regulatory regimes that either replace, manipulate, or completely silence conventional public administrative bodies (Daher 2008, Parnell and Robinson 2012, J. Sidaway 2007a, 2007b).

Experimental satellite cities

At the start of the 21st century, the accelerated pace of urbanization on a global scale has been accompanied by a wide variety of new experiments in city building. In what has become a familiar pattern throughout urban Africa, new large-scale master-planned cities built from scratch have begun to reshape existing metropolitan landscapes in ways unimaginable only a few decades ago. This shift towards fast-paced re-urbanism and away from incremental regeneration provides ample proof that urban Africa has not escaped this mania for experimentation. Real estate developers, business coalitions, and city officials have promoted these city-building efforts as 'world-class' urban spaces that contribute to the enhanced image of the city. Whether it is new master-planned enclaves carved out of derelict downtown spaces, new satellite cities built on the ex-urban fringe, or hyper-specialized enclaves that straddle major transportation corridors, these new city-building efforts seem to signal the beginning of the end of modernist thinking about rebuilding existing urban fabrics. Taken as a whole, they all point to innovative strategies designed to bypass broken-down infrastructure, overcrowded thoroughfares, and unworkable regulatory regimes.

While many of these urban redevelopment schemes amount to no more than the unrealized wishful thinking of overly optimistic property speculators, a number of large-scale real estate developers have already launched prototype mixed-use mega-projects on the outskirts of existing cities in Africa (De Boeck 2011a, Johnson 2013). Exemplary expressions of these ambitious re-urbanizing efforts in urban Africa include the massive shoreline reclamation project called Eko-Atlantic in Lagos; New Cairo City outside Cairo; high-tech business incubators like the $14.5bn Konza City technopolis, south of Nairobi; Machakos City, a satellite city south of Nairobi; King City, an exclusive mixed-use development near the rebuilt port of Takoradi (Ghana); Luano City, a satellite city built on the outskirts of Lubumbashi in cooper-rich Katanga Province (Democratic Republic of Congo); La Cité du Fleuve, a master-planned 'oasis of tranquillity' constructed on two islands reclaimed from sandbanks and swamp in the Congo River, adjacent to Kinshasa; Al Noor, a gateway city straddling Yemen and Djibouti and connected by the second longest bridge in the world; Malabo II in Equatorial Guinea; Steyn City, the egomaniacal fantasy-scape of the South African-born billionaire entrepreneur, Douw Steyn; and Waterfall City, an expansive, self-contained satellite city situated between Johannesburg and Pretoria (Johnson 2013). These new satellite cities unceremoniously inserted on the edges of existing metropolises mirror similar initiatives in the Asia Pacific Rim, the Persian Gulf, India, and the Middle East. Large-scale real estate developers and their state partners brand and market these 'neo-cities' as eco-friendly 'greenfield' projects that are functional, desirable, and sustainable alternatives to the growing slums, deteriorating and insufficient infrastructure, and inadequate housing typical of Africa's sprawling cities (Kihato and Kauri-Sebina 2012, Watson 2014).

Above all else, these large-scale city-building projects are designed to be 'self-sufficient' and functionally integrated, bringing together cutting-edge infrastructure, high-quality services, and opportunities for employment in safe-and-secure environments. The real estate developers who have spear-headed these new city-building projects and typically market them as smart, sustainable, and futuristic, combining a broad mixture of business oppor-tunities, luxurious residential accommodation, and leisure activities, as well as social amenities for their residents – from schools and medical centres to shopping malls, theatres and restaurants. These experiments in re-urbanism range from those well on their way to completion to those in the advanced stage of planning, and to those that have stalled due to lack of financing or poor forward planning. Besides those that amount to land speculation rip-offs and property profiteering schemes, there are those which are noth-ing more than the figments of fertile imaginations – the fantasy projections of dreamers and schemers (Choplin and Franck 2010, Watson 2014). Are these satellite cities only the latest iteration of master-planned modernist utopias at a time of late modernity, or are they actually realistic prototypes for city building in the 21st century? Do these mega-projects offer a presci-ent glimpse into the urban future of Africa, or are they just utopian fantasies that will never meet the enthusiastic expectations of over-confident real estate developers (Bhan 2014, Watson 2014)?

"Instant urbanism" and the reordering of urban temporality

Conventional city building involves patience and endurance, balancing piece-by-piece additions and subtractions with long time lines: cities grow, expand, and develop, or, conversely, stagnate, contract, and decline, in incremental steps and over lengthy intervals. In short, cities result from the cumulative, collective processes of everyday life and the complex interaction of all sorts of social forces. In this sense, cities are unpredictable, unruly, and virtually impossible to micromanage. In the contemporary age of globalization, the phenomenon of 'instant cities' (or 'pop-up cities' and fast-tracked urbanism) – with their particular kind of *tabula rasa* approach to city building – forces us to rethink what we mean by urbanism and urban life at a time of late modernity (Read 2005, MacDonald 2009, Moustafa et al. 2008; see also Barth 1975).

The accelerated pace of urbanization on a world scale has produced new metropolitan agglomerations at breakneck speed wherever real estate devel-opers have been able to seize a foothold in local property markets. This brand of fast-paced 'instant urbanism' is made visible in new 'model cities' and metropolitan agglomerations from the 'city-states' of the Persian Gulf to the emerging mega-cities of the Asia Pacific Rim, to the Special Economic Zones of India, to Songdo City outside Seoul in South Korea, and to Masdar in the United Arab Emirates, an eco-oasis in the desert, designed by 'starchitect' Norman Foster and built at a cost of $20bn for 40,000 inhabitants (Bach 2011, Dupont 2011, Goldman 2011, Shatkin and Vidyartha 2014, Wu 2007). These cities present themselves as operating at the forefront or cutting edge of the global frontier, as sustainable 'infrastructure cities' that unapologetically embrace utopian discourses of unrestrained entrepreneurialism and free market economics, in combination with fantasies of the clean slate that opens up new opportunities for innovative design and architectural experimentation (Barthel 2010, 133, Olds 2001, 1–10). In the metropolises of the Arab world, the construction of such mega-projects as *City of Silk* in Kuwait, *Al-Abdali* in Amman, *Solidere* in Beirut, and *Saphira* and *Bou Regreg* in Rabat reflects the experimental turn in city building towards the creation of stand-alone urban enclaves (Barthel 2010, 133).

While carefully and meticulously planned as showcase sites to attract global attention, these new cities are, willingly or not, also the material expressions of collective processes, complex collaborations, stealthy appropriations, and re-combinations of actually existing urban agglomerations. Despite their claims to novelty, they borrow, mimic, or steal ideas about building typologies, design aesthetics, and regulatory regimes at work elsewhere in the world economy (Schneekloth 2010). The construction of these new satellite cities is dependent upon 'travelling ideas', not freely circulating but moving along distinct pathways or channels traversed by global finance and adjusted to local circumstances (Tsing 2001, 13–35).

As a general rule, these new 'parallel cities' signify a kind of artificial 'ersatz urbanism'; that is, the design of urban spaces that resemble cities, but upon close inspection lack important elements which constitute how a 'real city' actually functions. Patterned on unimaginative modernist templates, real estate developers – via sophisticated advertising and promotional campaigns – have cleverly packaged these master-planned mega-projects as post-modern, avant-garde, and futuristic novelties. Often, the spatial design of these urban enclaves focuses on a narrow range of themed sites organized around functional specializations, while ignoring the kinds of genuine public spaces that foster chance encounter and civic engagement (de Wall 2007, 126). Constructed at unprecedented speed and on a colossal scale, these 'instant "cities-in-a-box"' (Boudreau 2010), as they have been called, have no recognizable central focal point, no genuine history, and no single or unifying identity except for 'newness' (Koolhaas 1998). Because they are built virtually overnight, they lack the kind of authentic sense of place that produces attachment and connection. It is sometimes difficult to think of these places as cities at all. Critics have frequently derided Dubai, which lays claim to some of the world's most expensive private islands, some of the tallest skyscrapers in the world, and perhaps the largest theme park, as a surreal playground where the very rich live in walled-off enclaves physically separated from the poor migrant workers who serve them (Bagaeen 2007, Davis 2006, Elsheshtawy 2010). In a similar vein, sceptics have criticized the awkward growth of Shenzhen (close to Hong Kong) as emblematic of what can happen when unregulated real estate development and property speculation are given free rein (Ouroussoff 2008, 71).

In contemporary European cities, part of the nostalgic attachment to historically preserved central squares and to assemblages of old buildings comes from a desire for permanence as a source of security at a time of uncertainty. Because of the sheer scale and the rapid speed at which they have been constructed, the fast-growing cities of China, India, and the Persian Gulf have very few of the features that architects and urban designers associate with a conventional modern metropolis (with its dense urban core and low-density residential suburbs). These mastered-planned new cities do not radiate from a historic centre as Paris, London, and New York do. Instead, their vast size and scale means that they function primarily as a series of decentralized zones and clustered nodal points, something closer in spirit to Los Angeles, Phoenix, and other fast-growing cities of the American west. The breathtaking speed of their construction means that they usually lack the layered mosaic of conventional city building – that is, the sometimes eclectic mixture of architectural styles and intricately related patchwork of socially stratified parts – that give older cities their complexity and distinctiveness: characteristics from which architects and designers have instinctively drawn inspiration. If New York represents the apotheosis of 19th- and 20th-century modernist city building, then Dubai, Abu Dhabi, and Doha (Qatar) symbolize the emerging prototype for the 21st century: prosthetic and detached oases presented as isolated satellite cities disconnected from what surrounds them (Bloch 2010, Kanna 2009, Mohammad and Sidaway 2012, Nagy 2000, Pacione 2005, Ponzini 2001, Shadid 2011).

In the wake of the retreat of public authorities from urban management and the consequent deterioration of the built environment, large-scale real estate developers have filled the void, assuming new planning powers once reserved for city officials that enable them to take the lead in reshaping the urban landscape. This approach to city building – which, following Gavin Shatkin (2011), might be called 'bypass-implant urbanism' – involves abandoning (or 'bypassing') congested and decaying spaces of the 'public city' to their own devices, and focusing instead on implanting large-scale, integrated mega-projects in strategic locations with easy access to up-to-date infrastructure. This new model of urban development is not merely a consequence of the blind adoption of 'Western' planning models, but reflects the incentives, constraints, and opportunities put into motion during the current phase of globalization (Shatkin 2008, 384). This approach to city building is largely dependent upon regressive state subsidies for profit-driven private sector development, the abandonment of the very idea of 'public purpose' and the 'common good' by municipal planning agencies, and the adoption of neoliberal modes of urban governance that idealize entrepreneurial solutions to city management (Graham 2000). Large-scale property developers do not just oversee the construction of integrated mega-projects; they take command over the conceptualization and implementation of entire infrastructural systems that are overlaid onto the existing metropolitan form. The kind of public–private partnerships that typically emerge reflect 'the domination of private developers in all planning processes, from the visioning of urban futures to the management of the urban environments that they create' (Shatkin 2008, 388). This model of urban development contrasts sharply modernist and high-modernist approaches with state-driven holistic planning and the public management of urban space (Harvey 1989, Jessop 1997, Murray 2013).

Building from scratch

Constructing these 'instant cities' out of whole cloth marks a decisive shift in city building away from the conventional planning approaches that typically focus on the revitalization or repair of existing urban landscapes. Starting afresh represents a new kind of holistic, rational planning for the contemporary urban age. As a curious blend of faith and science, these efforts at re-urbanism seem to resemble almost self-regulating, frictionless utopias that rest on the belief that 'smart city' designs and information technologies can produce technical solutions to the everyday problems of ordinary urban life (Datta 2015, Hollands 2008). The utopic stories that have accompanied the roll-out of these mega-projects are inextricably entangled with boosterist propaganda where optimistic visions of the future are virtually untethered from the nagging realities of everyday life in urban Africa. The new coalitions of real estate developers, finance capitalists, and design specialists have put their blind faith – what William Howard Kuntsler has called 'techno-grandiosity' or 'techno-triumphalism' – in the power of technology, the visionary prowess of entrepreneurialism, and the rule of experts (Kuntsler 2013, 12). They project a future urbanism without worry (Klingmann 2007, 1, 4, 6, 17–18, 35–36).

These re-urbanizing city-building projects intended to reshape the urban future of Africa depend upon deliberate strategic interventions that bypass existing metropolitan landscapes. Real estate developers, municipal authorities, and civic boosters of all sorts have legitimized the construction of these master-planned satellite cities as the much-needed 'solution' to the urban dysfunctionalities that severely hamper remedial efforts to rebuild or repair existing urban landscapes. Hailed by their promoters as cures or remedies for dysfunctional urbanism, these 'parallel cities' promise to deliver up-to-date infrastructure, good governance, and reliable services. The real estate developers engaged in these city-building projects are really in the business of finding spatial fixes to the social-technical challenges of urban Africa where most residents of existing large-scale metropolises live in unhealthy and unhygienic conditions, with little access to clean water, affordable housing, and regular wage-paying work (Simone 2001a, 2001b, 2003, 2004, 2005). The common denominator that these urban mega-projects share is a belief in the possibility of improving everyday life through expanded reliance on cutting-edge technology and up-to-date information systems.

Private cities: corporate take-over of city buildings

Without a strong civic authority at the local level to envision and oversee what is built and why, large-scale real estate developers with access to global financing and expert advice assume the commanding role in defining what kind of urbanity is created and sustained. Unburdened from the nagging need to comply with onerous standards or cumbersome regulations, property developers have free rein to build what they want and under conditions of their own choosing. Such commercially driven spatial enclaves as corporate business parks, innovation districts, free trade zones, and export-processing precincts are the primary aggregate units of a new species of contemporary city building, offering a clean slate, easy accessibility, and 'one-stop' entry into local economies (Easterling 2007a, 6; Easterling 2008), 75). The logistical requirements and organizational logics of global competitiveness provide generic specifications for assembling these new spatial envelopes. Contemporary versions of corporate business parks and other privately planned enclaves designed exclusively for global commerce are not merely alter-egos that attach themselves to the edges of existing cities, but often 'something more like independent city-states – the descendants of Venice or Genoa when they were the trading centers of the planet' (Easterling 2007a, 6: Easterling 2008).

Private cities have experimented with new forms of urban governance that effectively bypass conventional approaches to the public administration of urban space. The regulatory regimes that these satellite cities typically adopt are rooted in the entrepreneurial logic of expanding the capitalist marketplace to spheres of social activity that were once the exclusive preserve of the public administration. Private planning authorities composed of bureaucrats and technocrats have assumed widespread powers to buy, hold, and sell land and property; to manage bulk infrastructure (such as the street grid, storm water and sewage treatment, and sources of energy); to establish regulatory regimes with enforceable rules and mandatory compliance; and to oversee the provision of utilities like clean water, electricity, solar power, and natural gas (Sidaway 2003, 2007a, 2007b).

Implanted within existing sovereign bodies, these new satellite enclaves form political voids that function as territorial/legal loopholes in circuits of extraterritorial power (Franke and Weizman 2004, 4). By operating in this frictionless realm of legal ambiguity if not outright exemption, many of these newly minted 'city doubles' that have proliferated at the edges of existing cities in Africa find a relaxed atmosphere by hiding behind such spatial envelopes as Special Economic Zones (SEZs), Special Enterprise Zones, Free Trade Zones (FTZs) or Export Processing Zones (EPZs). As a distinct type of legal format, these 'zones of exception' (Agamben 2008) demarcate 'non-contiguous, differently administrated spaces of graduated or variegated sovereignty' (Ong 2006, 7).

While the nation-states within which these enclaves are embedded retain ultimate sovereignty, the private owners of these satellite cities maintain virtually complete jurisdiction and control for an indefinite period. In short, these satellite cities – as territorially demarcated enclaves – operate in what historian Elaine Scully has called 'anomalous zones', where conventional public administrative rules and norms are suspended and no longer apply. This legal anomaly is not an aberration, but instead is a defining characteristic of the extraterritorial governance that these urban enclaves need for their existence (Neuman 1996, Palan 2003, Scully 2001, 206 [n. 6]).

As the embodiment of calculated subterfuge, the enclave format provides the 'perfect legal habitat for corporate enterprise' (Easterling 2007b, 75). Aspiring to a state of exemption outside of the conventional functioning of the law and its restrictions, these commercially driven entrepôts seek to engineer their own status as self-sustaining islands of legal immunity and political quarantine specifically directed at corporate protection (Easterling 2007a, 10, 2005, 2). These various legal hybrids of the zone effectively oscillate between visibility and invisibility, and identity and anonymity. As 'off-worlds' disconnected from the seemingly chaotic street life of contemporary cities in Africa, these securitized enclaves provide a convenient platform enabling local property-holding elites to be both in and out of Africa at the same time (Easterling 2005, 48, 100, 116, 2008, 32).

The re-urbanizing efforts that have blossomed across urban Africa have come into existence under the sign of anti-statist neoliberal globalization. New satellite cities epitomize the worst excess of 'speculative urbanism' (Goldman 2011) – risky undertakings that depend in large measure on deliberate place-marketing and flashy branding. This kind of enclave urbanism signifies a pathway to hypermodernity, where building from scratch enables real estate developers to bypass existing urban landscapes and, metaphorically speaking, to leapfrog out of the past and propel themselves into the future (Shatkin 2008, 384). They represent the fantasy dreamscape of starting afresh, with their own rules and regulations and without outside encumbrances of poor municipal planning and public authority (Watson 2014). Some scholars have sought to classify these new city-building projects under the umbrella of an emergent model of *Dubaization* (Choplin and Franck 2010, 193). However, this analogy misses a crucial point; namely,

Dubai and the other 'instant cities' of the Persian Gulf emerged *de novo* in virtually uninhabited desert. In contrast, the new satellite cities of urban Africa have been inserted for the most part into existing urban agglomerations, thereby exacerbating spatial polarization (De Boeck 2011b, 74).

Working hand-in-glove with global design specialists and architecture firms, real estate developers with considerable financial backing have unveiled an ever-expanding number of ambitious plans for transforming urban Africa, promising that struggling mega-cities like Nairobi, Luanda, or Lagos – with chronic poverty, overcrowded streetscapes, and deteriorating infrastructures – can resemble Dubai, Singapore, or Shanghai. These new urban fantasies seek to circumvent if not erase the harsh realities of these cities, where the majority of urban residents remain extremely poor and are entrapped in informal housing with irregular work (Parnell and Pieterse 2010, Simone 2001a, 2001b, 2003, 2004, 2005). These new utopian fantasies are underpinned by the 'international best practice' rhetoric of 'smart cities', 'eco-cities', and sustainable urbanism. Yet these re-urbanizing projects are completely dependent upon siphoning off resources – water, labour, materials – from what surrounds them. In this sense, they are largely unsustainable in the extreme (Pieterse 2013, Pieterse and Parnell 2014, Watson 2014).

Typological classification

At the risk of simplification, it is possible to distinguish four distinct types of new city-building efforts in Africa. These can be identified primarily by their distinct functional specializations. The first type consists of large-scale mixed-use satellite cities that are typically constructed at the edge of existing urban landscapes and that operate under the mantra of 'live, work, and play'. The second type can be called 'terminal cities'; that is, new urban landscapes constructed around major inter-modal transportation hubs like ports, railroad lines, and airports. The third type can be labelled 'leisure-entertainment cities' which offer the amenities of a resort vacation in relaxed, outdoor settings. Finally, the fourth type consists of new 'information technology' hubs that function as business incubators.

Master-planned, multifunctional satellite cities: Eko-Atlantic, Lagos

Eko-Atlantic is a brand new city under construction on a seven-kilometre-long, two-kilometre-wide stretch of reclaimed coastline adjacent to Bar Beach, Victoria Island. This planned mixed-used property development is the centrepiece of a city redevelopment strategy whose ambition is indeed far-reaching: to remake the image of Lagos as an exemplar of distressed urbanism into Africa's model megacity. Described by the 1986 Nobel Prize winner Wole Soyinka as 'rising like Aphrodite from the foam of the

Atlantic', the Eko-Atlantic mega-project is perhaps the most ambitious land reclamation project ever undertaken in Africa.[2] Built in part to relieve the land pressure brought about by intense traffic congestion and overcrowded streetscapes, this entirely new satellite city is located on an estimated 1,000 hectares of entirely reclaimed land off Victoria Island. Promotional rhetoric has declared that this large-scale city-building project represents the 'Manhattan of urban Africa' and is destined to become the financial hub of the African continent.[3]

In the imagination of its private real estate developers, Eko-Atlantic is the mirror opposite of Lagos. Built entirely from scratch this master-planned mega-project sits atop an artificial island, built of sand and rock dredged from the ocean floor. A scale model on display at the offices of its property developers, South Energyx Nigeria, features tree-lined boulevards, crystal-clear canals, giant shopping malls, a well-designed waterfront with three marinas, a light-rail tramway, and a sail-shaped, 55-storey skyscraper that will be the new headquarters for a leading Nigerian bank. In promoting the project, David Frame, South Energyx Nigeria's managing director, hailed Eko-Atlantic as 'the new face of Africa'. Onno Ruhl, country head for the World Bank, proclaimed this mega-project as 'the future Hong Kong of Africa'.[4]

The private real estate developers behind the Eko-Atlantic project have declared that their ultimate goal is to 'build a self-contained city capable of providing the quality of infrastructure and support services required to transform Lagos into an economic powerhouse and the financial hub of Africa' (see Murray 2014, 101). Spatial plans for this nine-square-kilometre (four square miles) redevelopment project envision six distinct components, including a central business district, waterfront leisure-entertainment sites, and residential accommodation – and these are all connected by efficient transport systems. The design specialists who conceived of this new satellite city have promised that each district will include mixed-use spaces that combine high-end residential developments with business offices and commercial buildings, leisure-and-entertainment facilities, and retail shopping.[5]

Eko-Atlantic is designed to provide residential accommodation for upwards to half a million people with 250,000 daily commuters, and office towers and other workplaces for another 150,000. In planning for this City of the Future, the real estate developers behind the massive project have included high-tech infrastructure in line with environmental standards defined by international best practice. Upon completion, Eko-Atlantic will offer its residents state-of-the-art waste management and surface water drainage, clean water supply and distribution, security and transportation systems, along with information technology networks capable of serving a world-class central business district. This 'city-within-a-city' will also provide its residents with an independent source of energy provided by an electrical power generation plant with underground distribution and the capacity to provide for the entire complex (Clayton 2011).

Undeterred by concerns about future vulnerability brought about by rising sea levels in the coming century, real estate developers have pressed ahead. With land reclamation virtually complete, the main private real estate developers – the Chagoury family – have embarked on the second phase: the layout of the spatial grid installation of bulk infrastructure. While Eko-Atlantic is formally a public–private partnership between the Lagos State Government and South Energyx (recipient of the Clinton Global Initiative Commitment Certificate in 2009), its spatial design, overall functions, and systems of urban management are virtually entirely in the hands of private real estate companies (Hinshaw 2013).[6]

Transportation and logistics hubs:
Minna Airport City

While the mega-project was still in the visionary stage, real estate developers have drawn up ambitious plans to transform Minna, the capital city of Niger State in northern Nigeria, into an agro-industrial, manufacturing, aviation logistics centre built around an existing local airport. Borrowing freely from existing 'airport city' mega-projects in Amsterdam, Dallas/Fort Worth, Dubai, Dusseldorf, Frankfurt, Vienna, Hong Kong, Kuala Lumpur, and Singapore, the large-scale property developers behind the Minna Airport City (MAC) scheme envision building an entirely new city on a sprawling 30,000-hectare site, complete with a supermarket of facilities, including state-of-the-art business offices, recreational components, upscale shopping malls, light manufacturing, health industries, educational facilities, and residential housing estates that could accommodate six million residents.

The master plan for the MAC project includes a dedicated zone for light manufacturing, cargo warehousing facilities, and a logistics centre specializing in the handling and export of agricultural products. The main stakeholders in the MAC scheme have hoped to use the development model of a public–private partnership as the catalyst for jump-starting their long-term goal of redefining the entire sub-region of West Africa through the creation of a transportation hub linking nearby Abuja with the Baro port on the coast and manufacturing centre capable of undercutting cheap imports, mainly from China.[7]

Themed entertainment destinations: Casa Marina
and Mall of Morocco (Casablanca)

City builders in Morocco have aggressively tried to fashion an image of its new city-building projects as the Dubai of the Maghreb. Lacking the revenue of resource-rich Gulf States, real estate developers have turned to European loans and investments to finance their own brand of 'petro-urbanism'.

The flagship 'mega-project' is the $500-million redevelopment of the underutilized Casablanca waterfront into the luxurious, mixed-use Casa Marina. Stretching across 450,000 square meters, the complex will include high-end retail spaces, an integrated business centre and 'luxury apartments targeted at VIP customers'. The Central Real Estate Company (CGI) has announced its intention to add three artificial islands to expand the size of the Marina. Prestige mega-projects like the Marina are very powerful frameworks for the design and implementation of complex urban operations where private real estate interests are the driving force. Moreover, it would seem that major tensions between the common interest and the profit imperative in some ways weaken this new form of global spatial production. Beyond the official discourses, prestige projects have taken territorial governance in new directions, involving multi-scale public–private partnerships that seek to circumvent genuine participatory approaches (Barthel and Planel 2010, Bogaert 2012, Zemni and Bogaert 2011).

City boosters hailed the opening in December 2011 of the Mall of Morocco in Casablanca as the first world-class 'destination' shopping mall in Africa, and one of five of its kind in the world. With an inauguration ceremony featuring a controversial lip-synched performance by pop-star Jennifer Lopez, the sprawling mall complex – what insiders have called a 'retailtainment' mecca – consists of around 350 stores, 40 gourmet restaurants, a giant aquarium, an IMAX 3D theatre, an amusement park, a VIP shopping area, a large aquarium, an ice skating rink, a bowling centre, a musical fountain, lush gardens, and meditation pads. Designed to resemble a giant seashell, the architectural design of the mega-project blurs the line between a crass imitation of Western modernity imitation and historical preservation of the traditional souk marketplace. The interior design of the department store Galeries Lafayette, which first opened its doors in Casablanca during French colonial rule in the 1920s but was later closed after independence, mimics the Galeries Lafayette in Paris. The musical fountain outside the back side of the mall facing the ocean bears a striking resemblance to a similar water-and-music synchronized display at the Las Vegas Bellagio Hotel. The glossy advertisements that overwhelm the surrounding streetscape, and the hundreds of European and American retail franchises that fill the halls inside, demonstrate the desire to instil Western consumption habits in a social world still grappling with postcolonial development. Behind this new symbol of luxury lies a complex web of compromised financial interests and a vision of profit-making enterprise that seems to neglect the yawning gap between rich and poor. While a projected 14 million visitors will enjoy the upscale shopping and exclusive spas at the largest shopping mall in Africa, urban residents of the surrounding neighbourhoods continue to live in poverty (Berry-Chikhaoui 2010, Bogaert 2013).

Themed retail destinations: Mall of Mauritius at Bagatelle and the La Balise Marina (Mauritius)

Over the past decade, the small island nation of Mauritius has positioned itself as a prominent offshore financial-banking centre that prides itself on secrecy and confidentiality. In the World Bank's 2009 Doing Business report, it was ranked 24th in the world and first in Africa. It was also rated among the top 25 global offshore platforms. With the number of globe-trotting tourists expected to reach two million by 2015, Johannesburg-based Atterbury Developments, a leading property investment and development company, teamed with ENL Property, a well-established Mauritian real estate holding company with strong ties to South African firms, to finance and manage the construction of the Mall of Mauritius at Bagatelle. By creating a carefully crafted place-image of 'an authentic tropical "sense of ease"', corporate boosters have promoted this multifunctional shopping mall as a shopping destination like 'no other in the Indian Ocean', where 'enthusiastic shopaholics will be spoilt for choice as they match and accessorize to their heart's content'. Outfitted with the large collection of globally recognized designer brands under one roof, city boosters have lauded the Mall of Mauritius at Bagatelle as 'the next fashion and entertainment capital of island living'.[8]

Like many of the projects in the ELN Property portfolio, the La Balise Marina project (located along the Black River in close proximity to the Mall of Mauritius at Bagatelle) falls under the property development guidelines of an 'Integrated Resort Scheme development', a government-sponsored initiative aimed entirely at attracting wealthy foreigners who wish to invest in high-end residential accommodation, offering them a vacation home in an island paradise and automatic permanent residency status. This large-scale property development offers a marina lifestyle built around a 'sea and sun experience'. Promotional materials refer to the La Balise Marina as an idyllic place offering 'incomparable outdoor lifestyle' choices that blend yachting and boating, fishing and ocean-cruising, snorkelling and water skiing. In promoting the unique qualities of these sun-and-sea dreamscapes, sales representatives stress the idyllic landscape of sandy beaches, tropical breezes, and blue skies.[9]

Privately built and privately maintained places like the Mall of Mauritius at Bagatelle and the La Balise Marina illustrate this global trend towards combining international corporate headquarters with a full range of cultural, leisure, and entertainment opportunities in a resort-like atmosphere. Ambitious mixed-use developments featuring luxury residences, office blocks, hotels, huge shopping malls, and imaginative entertainment complexes have fundamentally altered the image of Mauritius and put the local property market on the world stage. On a broader scale, Atterbury Developments, in collaboration with ENL Property, is the driving force behind the Bagatelle commercial development zone on the outskirts of the capital Port Louis.

The Bagatelle integrated development project is a self-contained 'mini-city' consisting of warehousing and light industrial facilities, business office parks, the Mall of Mauritius, the Voila Hotel, leisure-and-entertainment venues, and luxury residences designed for affluent homeowners, 'interlinked with a strong synergy in a mixed-use environment'.[10] The long-term vision for the Bagatelle commercial development zone is to create a thriving and aesthetically pleasing, mixed-use precinct that functions as a regional hub of global commerce, a high-profile tourist destination, and a shopping mecca.[11]

ENL Property is a multifunctional conglomerate involved in multiple projects in Mauritius, particularly in the agribusiness, commercial, and property sectors. In addition to the La Balise Marina, ENL Property is engaged in a variety of real estate projects, including Villas Valriche, a luxury residential development located in the south of the island; Moka, Le Coeur de l'ile, a mixed-use project consisting of a luxury residential development, sports/shopping centre, and business park, located in the Moka mountains; and Kendra, a commercial hub situated near St Pierre.[12]

Incorporated in 2005 as a regional subsidiary of the state-owned conglomerate Dubai World Corporation (Dubai's flag bearer in global investments), Dubai World Africa specializes in leisure-and-entertainment properties, wildlife conservation, and eco-tourism in Africa. The company has property investments in high-end real estate resort projects, hotels, and luxury lodge developments in South Africa, Mozambique, Comoros, Zanzibar, Djibouti, and Rwanda, with plans (partially put on hold following the worldwide financial downturn of 2008) to expand its tentacles into Morocco, Senegal, Gabon, and Benin. In 2006, it acquired (but subsequently sold in 2011) the iconic Victoria & Alfred Waterfront in Cape Town. In 2007, it purchased Daallo Airlines, the state-owned air carrier of Djibouti, in an effort to further facilitate the growth of trade and tourism on the African continent (Anonymous 2008). Dubai World Africa has embarked on an ambitious strategy to rehabilitate, develop, and manage high-end game parks and wildlife reserves across the African continent to tap into the accelerating demand for eco-tourism and adventure tourism. Its portfolio of African properties includes the Pearl Valley Signature Golf Estate and Spa (Cape Town); Shamwari Game Reserve, Sanbona Wildlife Reserve and Jock Safari Lodge (South Africa); and the Kempinski Luxury Beach Resort in the Comoros, One & Only Zanzibar, and Djibouti Palace Kempinski. In Rwanda alone, Dubai World Africa owns such high-end properties as the Akagera Safari Lodge, Gorilla's Nest Mountain Lodge, Nyungwe Forest Lodge, and the Kigali Residence, Hotel and Golf Club. Inspired by steadily rising trade between Dubai and the Republic of Benin, Dubai World Africa has entered into negotiations with local investors to construct a 32-hectare master-planned beachfront development along with an international hotel in the capital city of Porto Novo. The aim of these property investments is to provide a gateway into the potentially lucrative business of eco-tourism and adventure tourism.[13]

Hi-tech cities: Hope City, Accra (Ghana) and Kakungulu Satellite City (Kampala, Uganda)

Launched with great fanfare in 2013, construction has stalled at Hope City, a multi-faceted $10-billion high-tech hub located on a sprawling 1.5-million-square-meter site roughly 30 kilometres west of Accra. Once expected to take three years to complete, this ambitious project is the brainchild of Ghanaian businessman Roland Agambire, head of local technology group RLG Communications. Real estate developers (particularly the AGAMS Group) seek to use this new satellite city to transform Ghana into a major player in competition with the nascent information communications technology (ICT) industry emerging in East Africa. The architects who designed this satellite city have sought to construct an integrated living and working space that functions as a habitat for ICT industries engaged in product research, design, and manufacture. Smart and futuristic, this proposed high-tech city includes a large assembly plant for various technology products, business offices, and an IT university.[14]

Designed by Italian firm Architect OBR, the vision for this planned technopolis includes a Dubai-style skyline consisting of six towers of different dimensions, clustered around an iconic 75-storey, 270-meter-high building that is expected to be the highest in Africa.[15] Bridges of varying heights will connect various districts of the new city, including residential accommodation for around 25,000 residents, a state-of-the-art private hospital, recreational spaces, and other social amenities such as restaurants, theatre complexes and leisure-entertainment activities. The telecom giant RGL Communications has spearheaded the initial stages of this mega-project, with funding to cover around 30 per cent of the overall costs, while the remainder of the financing has come from a wide array of private investors and a stock-buying scheme. Acting in partnership with private capital, the Ghanaian government has used various incentives at its disposal to attract well-known companies in the global ICT industry, including Microsoft, to establish offices in this research-and-development hub, which is expected to create jobs for an estimated 50,000 people.[16]

On the green, thousand-acre hillside on the Kampala-Entebbe highway in south-central Uganda, real estate developers have already started work on a new city that is projected to turn the country into the African continent's answer to Bangalore, the main IT hub of India.[17] The centrepiece of the large business estate consists of a vast IT park, including call centres, back-office facilities, and data-processing operations – all of which will be much cheaper for multinational firms to operate than those in Britain. Owned and managed by Akright Projects, Kakungulu Satellite City is also designed to include a central business district, a 50,000-seater sports stadium, a hospital, exclusive housing developments, a luxury hotel, two shopping malls, schools, and a signature golf course – with seeds for the greens flown in from

Florida (Johnson 2013). The residential precinct contains 2,500 dwellings and is expected to be home to nearly 144,000 people. Kakungulu Satellite City is set to employ a huge pool of Ugandan graduates in outsourced Western-style call centres. Real estate developers behind this new experiment in 'organized living' (as promotional materials promise) expect it to generate a projected 20 per cent of the total of the yearly exports of Uganda in just five years.[18]

The future of urbanism in Africa

What these satellite cities share is the fantasy-scape of offering luxury lifestyles in urban Africa. By abandoning the goal of making improvements to existing urban landscapes, these master-planned enclaves present themselves as an alternative to mega-urbanization in Africa.

The scope and scale at which these efforts at re-urbanism are taking place in Africa, and the loud proclamations of real estate developers and public authorities regarding the planned transformation of urban life, have produced a varied and strong response from both urban scholars and local policy-makers. In this context, both popular and academic accounts of the current state of new city-building projects in urban Africa quite often read like interpretations of a Rorschach test – they may say more about the hopes and anxieties of the observer than about the images themselves (Shatkin 2014, 11). On one side, the real estate developers behind these mega-projects – along with their coteries of place-promoters, boosters, and ideological advocates – see these new satellite cities as a prescient glimpse of a hopeful future of a fully modernized and up-to-date metropolis (de Boeck 2011b). On the other side, sceptics suggest that these large-scale projects 'portent of a future slow-motion apocalypse', where the urban poor are pushed aside and 'brutally dislocated to make way for a partitioned city in which the elite retreat into fortified enclaves' (Shatkin 2014, 11). What is needed is a sanguine assessment of the evolving dynamics of city building in Africa, to not only gauge the direction and impact of these changes, but also contribute to debates about how urban planners, policy-makers and community organizers might intervene to shape what is happening (Murray 2013, 204; Shatkins 2014, 11; Watson 2014).

Pushed to their limits, these satellite cities amount to little more than large-scale infrastructure projects, targeted at corporate business and affluent residents. They function as autonomous 'city-states' that operate by a different set of formal and tacit rules and regulations from the surrounding hinterland.

The juxtaposition between these planned satellite cities and current state of affairs in urban Africa reveals the tensions that exist between the utopian dream of starting over and the failure of planners and policy-makers to develop realistic remedies to incorporate ordinary residents into the mainstream of urban life. The fundamental disjuncture between the here-and-now of actual urban life and the there-and-then of sweeping gestures of future utopias has reinforced a false opposition that substitutes wishful thinking for the harsh realities of everyday life in the African city (de Boeck 2011b, 74, 82). As Edgar Pieterse (2008, 4–6, 84–87) has argued, the existential core of urbanism is the desire for genuinely radical transformation, but this radical impulse often stands in sharp contrast to the kind of 'necessary prudence' which characterizes a more incremental, bottom-up approach to the pressing problems of everyday life that often seems to be lacking in cities in Africa. It is this profound disjuncture between 'the on-the-ground practices of everyday urban life and survival' and the elitist urge to start over with new satellite cities that produces the utopian dreamscape of escapist urbanism (de Boeck 2011b, 74) – the fantasy that technology, infrastructure, good governance, and enforceable rules offer a 'quick fix' that can make the present state of affairs simply disappear – evaporate as if it never existed (Johnson 2013; Murray 2013, 204; Watson 2014).

Notes

1 Some of the ideas contained in these first several paragraphs are borrowed from Murray (2014).
2 Wole Soyinka, "Wole Soyinka on Lagos," *Newsweek,* 6 March 2011.
3 Alex Perry, "Making Over Lagos," *Time,* 26 May 2011; Jonathan Clayton, "Eko Atlantic: Africa's City of Tomorrow," *The Times of London,* 2 May 2011; and Drew Hinshaw, "Nigerian Developer Set to Build Africa's Next Giant City," *Wall Street Journal,* 12 August 2013.
4 Two quotations taken from Egon Cossou, "Lagos Aims to be Africa's Model Megacity," *Africa Business Report, BBC World News,* 22 January 2010. See also Interview with Koome Gikunda, Investment Principal, Actis (Nairobi, 15 August 2012).
5 Egon Cossou, "Lagos Aims to be Africa's Model Megacity," *Africa Business Report, BBC World News,* 22 January 2010.
6 Femi Akinola, "A Fairy Tale, Dream Home for the Super-Rich," *Sunday Trust* (Abuja), 19 December 2010.
7 Nahimah Ajianle Nurudeen, "Minna to Have First Airport City in Africa," *Daily Trust* (Abuja), 30 March 2010; Nahimah Ajikanle Nurudeen, "Nigeria: Minna Airport City Will Grow Our Economy," *Daily Trust* (Abuja), 14 April 2010; and S. O. J. Bamidele, "Minna Airport City – Will Aliyu's Dreams Come True?" *Daily Champion,* 17 March 2011.
8 For the source of quotations, see "SA Developer Partners in Mauritius Retail," *EPROP Commercial Marketplace,* 13 October 2009. See Dennis Ndala, "SA Company Developing Mauritius' First Shopping Mall," *Engineering News,* 22 June 2011.
9 See "South Africans Keen on Mauritian Marina," *Business Day,* 30 September 2009. See also La Balise Marina, *The United Nations Development Project Studies La Balise Marina's Social Integration Project* (2012) [Retrieved from La Balise Marina: Black River, Mauritius: <http://www.labalisemarina.com/en/news/undp.aspx>]; and La Balise Marina, *Vision* (2012) [Retrieved from La Balise Marina: Black River, Mauritius: http://www.labalisemarina.com/en/project/vision.aspx].

10 "Strong Support as Atterbury Develops Mall of Mauritius," *Business Day,* 20 July 2011 (p. 8).
11 ENL Property, *Our Projects* (2009) [Retrieved from ENL Property: <http://www.enl.mu/en/enl_property/our_projects.aspx>]. See also Geneva Management Group, *Mauritius: High-level overview of the Integrated Resort Scheme ('IRS') and Real Estate Scheme ('RES')* (2010) [Retrieved from Geneva Management Group <http://www.gmgtrust.com/High-Level-overview>].
12 See ENL Property, *Our Projects* (2009) [Retrieved from ENL Property: <http://www.enl.mu/en/enl_property/our_projects.aspx>].
13 Dubai World Africa, "Dubai World Prospects Investment Possibilities at Port Cotonou with Republic of Benin," *EPROP Commercial Marketplace,* 18 April 2008.
14 See Teo Kermeliotis, "'Africa's Tallest Building' set for $10 billion Tech City," *CNN Marketplace Africa,* 20 March 2013.
15 See Teo Kermeliotis, "'Africa's Tallest Building' set for $10 billion Tech City," *CNN Marketplace Africa,* 20 March 2013.
16 U.S. Chamber of Commerce, International Affairs, Africa Business Initiative, *Investment Climate Update: Techno Cities* 4, 1 (September 2013), pp. 1–4.
17 The choice of name is quite interesting. Kakungulu was a Ganda chief in the 19th century who abjured medicine and inoculations, broke away from the Ganda hierarchy, and set himself up as chief in Busoga, in the east, where he founded a Ugandan Jewish sect. That this island-like techno-paradise would be named after an anti-technology dissenter is indeed bizarre. (Thanks to Derek Peterson, Department of History, University of Michigan, for this insight.)
18 Sitinga Kachipande, "A New African Dream: The Benefits of Satellite Cities for East Africa," *Consultancy Africa Intelligence,* 16 May 2013. Greater Horn of Eastern Africa (GHEA) outlook #18: "Are Satellite Cities the (official) Future of GHEA's Urbanisation?" Society for International Development, 2010. See also Akright Development website available at <http://www.akright.biz>.

References

Agamben, G. 2008. *State of Exception.* Chicago: University of Chicago Press.
Anonymous. 2008. "Dubai world gives African tourism a boost." *Financial Mail,* January 11.
Bach, J. 2011. "Schenzen: city of suspended possibility." *International Journal of Urban and Regional Research* 35(2): 414–420.
Bagaeen, S. 2007. "Brand Dubai: the instant city or the instantly recognizable city." *International Planning Studies* 12(2): 173–197.
Barth, G. 1975. *Instant Cities: Urbanization and the Rise of San Francisco and Denver.* New York: Oxford University Press.
Barthel, P-A. 2010. "Arab mega-projects: between the Dubai effect, global crises, social mobilization and a sustainable shift." *Built Environment* 36(2): 133–145.
Barthel, P-A. and Planel, S. 2010. "Tanger-Med and Casa-Marina, prestige projects in Morocco: new capitalist frameworks and local context." *Built Environment* 36(2): 176–191.
Berry-Chikhaoui, I. 2010. "Major urban projects and the people affected: the case of Casablanca's Avenue Royale." *Built Environment* 36(2): 216–229.
Bhan, G. 2014. "The real lives of urban fantasies." *Environment and Urbanization* 26(1): 1–4.

Bloch, R. 2010. "Dubai's Long Goodbye." *International Journal of Urban and Regional Research* 34(4): 943–951.

Bogaert, K. 2012. "New state space formation in Morocco: the example of the Bouregreg Valley." *Urban Studies* 49(2): 255–270.

Bogaert, K. 2013. "Cities without slums in Morocco? New modalities of urban government and the Bidonville as neoliberal assemblage." In *Locating the Right to City in the Global South*, by T. Samara, S. He and G. Chen (eds.), 41–59. New York: Routledge.

Boudreau, J. 2010. "CISCO systems helps build prototype for instant 'city in box'." *Washington Post*, June 9.

Choplin, A. and Franck, A. 2010. "A glimpse of Dubai in Khartoum and Nouakchott: prestige urban projects on margins of the Arab world." *Built Environment* 36(2): 192–205.

Clayton, J. 2011. "Eko Atlantic: Africa's city of tomorrow." *The Times of London*, May 2.

Daher, R.F. 2008. "Amman: disguised genealogy and recent urban restructuring and neoliberal threats." Chap. 3 in *The Evolving Arab City: Tradition, Modernity and Urban Development*, by Y. Elsheshtawy (ed.), 37–63. New York: Routledge.

Datta, A. 2015. "New urban utopias of postcolonial India: entrepreneurial urbanization in Dholera smart city, Gujarat" [Anchor paper]. *Dialogues in Human Geography* 5(1): 3–22.

Davis, M. 2006. "Fear and money in Dubai." *New Left Review* 41: 47–68.

De Boeck, F. 2011a. "Inhabiting ocular ground: Kinshasa's future in the light of Congo's spectral urban politics." *Cultural Anthropology* 26(2): 263–286.

De Boeck, F. 2011b. "The modern titanic: urban planning and everyday life in Kinshasa." *The Salon (Johannesburg Workshop in Theory and Criticism* 4: 73–82.

De Boeck, F. 2012. "Spectral Kinshasa: building a city through the architecture of words." In *Urban Theory Beyond the West: A World of Cities*, by T. Edensor and M. Jayne (eds.). New York: Routledge.

De Wall, M. 2007. "Power urbanisms." In *Visionary Power: Producing the Contemporary City*, by C. de Baan, J. Declerck and V. Patteeuw. Rotterdam: NAi Publishers.

Dupont, V. 2011. "The dream of Delhi as a global city." *International Journal of Urban and Regional Research* 35(3): 533–554.

Easterling, K. 2005. *Enduring Innocence: Global Architecture and its Political Masquerades*. Cambridge, MA: MIT Press.

Easterling, K. 2007a. "Extrastatecraft." In *Perspecta 39: Re-Urbanism: Transforming Capitals*, by K. Agarwal, M. Domino, E. Richardson and B. Walters (eds.). Cambridge, MA: MIT Press.

Easterling, K. 2007b. "The zone." In *Visionary Power: Producing the Contemporary City*, by de C. Baan, J. Declerk and V. Patteeuw (eds.). Rotterdam: NAi Publishers.

Easterling, K. 2008. "Zone." In *Urban Transformation*, by I. Ruby and A. Ruby (eds.). Berlin: Ruby Press.

Elsheshtawy, Y. 2010. *Dubai: Behind an Urban Spectacle*. New York: Routledge.

Franke, A. and Weizman, E. 2004. "The geography of extraterritoriality." In *Territories: The Frontiers of Utopia and Other Facts on The Ground*, by A. Franke and E. Weizman. Koln, Germany: Verlag der Buchhandlung Walther Konig.

Goldman, M. 2011. "Speculative urbanism and the making of the next world city." *International Journal of Urbana and Regional Research* 35(3): 555–581.

Graham, S. 2000. "Constructing premium network spaces: reflections on infrastructure networks and contemporary urban development." *International Journal of Urban and Regional Research* 24(1): 183–200.

Harvey, D. 1989. "From managerialism to entrepreneurialism: the transformation of urban governance in late capitalism." *Geografiska Annaler* 71B: 3–17.

Hinshaw, D. 2013. "Nigerian developer set to build Africa's next giant city." *Wall Street Journal*, August 12.

Hollands, R. 2008. "Will the real smart city please stand up?" *City* 12(3): 303–320.

Jessop, B. 1997. "The entrepreneurial city: re-imagining localities, redesigning economic governance, or restructuring capital?" In *Transforming Cities: Contested Governance and New Spatial Divisions*, by N. Jewson and S. MacGreggor (eds.). New York: Routledge.

Johnson, L. 2013. "Petropolis now: are cities getting too big?" *New Statesman*, 14 November 2013. http://www.newstatesman.com/writers/116085.

Kanna, A. 2009. *The Superlative City: Dubai and the Urban Condition in the Early Twenty-First Century*. Cambridge, MA: Harvard University Graduate School of Design.

Kihato, M. and Kauri-Sebina, G. 2012. "Urban soci-spatial change and sustainable development: the neo-city phenomenon." In *Perspectives: Political Analysis and Commentary from Africa#3*, by C. Peter (ed.). Capetown: Heinrich Boll Stiftung.

Klingmann, A. 2007. *Brandscapes: Architecture and the Experience Economy*. Cambridge, MA: MIT Press.

Koolhaas, R. 1998. "The generic city." In *S,M,L,XL*, by R. Koolhaas and B. Mau (eds.). New York: Monacelli Press.

Kuntsler, W.H. 2013. *Too Much Magic: Wishful Thinking, Technology, and the Fate of the Nation*. New York: Grove Press.

Lumumba, J. 2013. "Why Africa should be wary of its 'new cities'." *Informal City Dialogues (Rockfeller Center)*. May 2. Accessed February 5, 2014. https://nextcity.org/informalcity/entry/why-africa-should-be-wary-of-its-new-cities.

MacDonald, M.-P. 2009. "Pop-up cities." *ETC* 87: 18–24.

McCann, E. and Ward, K. 2011. "Urban assemblages: territories, relations, practices and power." In *Mobile Urbanism: Cities and Policymaking in the Global Age*, by E. McCann and K. Ward (eds.). Minneapolis: University of Minnesota Press.

Mohammad, R. and Sidaway, J. 2012. "Spectacular urbanization amidst variegated geographies of globalization: learning from Abu Dhabi's trajectory through the lives of South Asian men." *International Journal of Urban and Regional Research* 36(3): 606–627.

Moustafa, A., Al-Qawasami, J. and Mitchell, K. (eds.). 2008. "Instant cities: emergent trends in architecture and urbanism in the Arab world." *The Third International Conference of the Center for Study of Architecture in the Arab Region*. Sharaj, United Arab Emirates: CSAAR Press.

Murray, M. 2013. "Afterword: re-engaging with transnational urbanism." In *Locating Right to City in the Global South*, by T.R. Samara, S. He and G. Chen (eds.). London: Routledge.

Murray, M. 2014. "City doubles: re-urbanism in Africa." In *Cities and Inequalities in a Transnational World*, by F. Miraftab, K. Salo and D. Wilson (eds.). New York: Routledge.

Myers, G. and Murray, M. 2006. "Introduction: situating contemporary cities in Africa." In *Cities in Contemporary Africa*, by M. Murray and G. Myers (eds.). New York: Palgrave.

Nagy, S. 2000. "Dressing up downtown: urban development and government public image in Qatar." *City and Society* 12(1): 125–147.

Neuman, G. 1996. "Surveying law and borders: anomalous zones." *Stanford Law Review* 48(5): 1197–1234.

Olds, K. 2001. *Globalization and Urban Change: Capital, and Pacific Rim Mega-Projects*. Oxford: Oxford University Press.

Ong, A. 2006. *Neoliberalism as Exception: Mutations in Citizenship and Sovereignty*. Durham, NC: Duke University Press.

Ouroussoff, N. 2008. "The new, new city." *New York Times Magazine*, June 8.

Pacione, M. 2005. "Dubai." *Cities* 22(3): 255–265.

Palan, R. 2004. *The Offshore World: Sovereign Markets, Virtual Places, and Nomad Millionaires*. Ithaca: Cornell University Press.

Parnell, S. and Pieterse, E. 2010. "The 'right to the city': institutional imperatives of a developmental state." *International Journal of Urban and Regional Research* 34(1): 146–162.

Parnell, S. and Robinson, J. 2012. "(Re)theorizing cities from the global south: looking beyond neoliberalism." *Urban Geography* 33(4): 593–617.

Pieterse, E. 2008. *City Futures: Confronting the Crises of Urban Development*. London: Zed Press.

Pieterse, E. and Simone, A.M. (eds.). 2013. *Rogue Urbanism: Emergent African Cities*. Cape Town: Jacana.

Pieterse, E. and Parnell, S. (eds.). 2014. *Africa's Urban Revolution*. Cape Town: UCT Press.

Ponzini, D. 2011. "Large scale development projects and star architecture in the absence of democratic politics: the case of Abu Dhabi." *Cities* 28(3): 251–259.

Read, S. 2005. "The form of the future." In *Future City*, by S. Read, J. Roseman and J. van Eldijk (eds.). London and New York: Spon Press.

Roy, A. 2011. "Slumdog cities: rethinking subaltern urbanism." *International Journal of Urban and Regional Research* 35(2): 223–238.

Schneekloth, L. 2010. "Conclusion: seeing through: types and the making and unmaking of the world." In *Re-Shaping Cities: How Global Mobility Transforms Architecture and Urban Form*, by M. Guggenhem and O. Soderstrom (eds.). New York: Routledge.

Scully, E. 2001. *Bargaining with the State from Afar: American Treaty Citizenship in Port China, 1844–1942*. New York: Columbia University Press.

Shadid, A. 2011. "Qatar's capital glitters like a world city, but few feel at home." *New York Times*, November 30.

Shatkin, G. and Vidyartha, S. 2014. "Contesting the Indian city: global visions and the politics of the local." In *Contesting the Indian City: Global Visions and the Politics of the Local*, by G. Shatkin (ed.). London and New York: Wiley Blackwell.

Shatkin, G. 2008. "The city and the bottom line: urban megaprojects and the privatization of planning in Southeast Asia." *Environment and Planning* 40: 383–401.

Shatkin, G. 2011. "Planning privatopolis: representation and contestation in development of urban integrated mega-projects." In *Worlding Cities: Asian Experiments and the Art of Being Global,* by A. Roy and A. Ong (eds.). Malden, MA: Blackwell.

Shatkin, G. 2014. "Contesting the Indian city: global visions and the politics of the local." *International Journal of Urban and Regional Research* 38(1).

Sidaway, J. 2003. "Sovereign excesses? Portraying postcolonial sovereigntyscapes." *Political Geography* 22: 157–178.

Sidaway, J. 2007a. "Enclave space: a new metageography of development?" *Area* 39(3): 331–339.

Sidaway, J. 2007b. "Spaces of postdevelopment." *Progress in Human Geography* 31(3): 345–361.

Sigler, T. 2013. "Relational cities: Doha, Panama City, and Dubai as 21st century entrepots." *Urban Geography* 34(5): 612–633.

Simone, A-M. 2001a. "Straddling the divides: remaking associational life in the informal African city." *International Journal of Urban and Regional Research* 25(1): 102–117.

Simone, A-M. 2001b. "On the worlding of African cities." *African Studies Review* 44(2): 15–41.

Simone, A-M. 2003. "Resources of intersection: remaking social collaboration in urban Africa." *Canadian Journal of African Studies* 37(2–3): 513–538.

Simone, A-M. 2004. *For the City yet to Come.* Durham, NC: Duke University Press.

Simone, A-M. 2005. "People as infrastructure: intersecting fragments in Johannesburg." *Public Culture* 16(3): 407–429.

Smith, N. 2002. "New globalism, new urbanism: gentrification as global urban strategy." *Antipode* 34(3): 427–450.

Tsing, A. 2001. *Friction: An Ethnography of Global Connection.* Princeton, NJ: Princeton University Press.

Watson, V. 2009. "Seeing from the south: refocusing urban planning on the globe's central urban studies." *Urban Studies* 11: 2259–2275.

Watson, V. 2014. "African urban fantasies: dreams or nightmares?" *Environment and Urbanization* 26(1): 215–231.

Wu, F. (ed.). 2007. *China's Emerging Cities: The Making of New Urbanism.* New York: Routledge.

Zemni, S. and Boagert, K. 2011. "Urban renewal and social development in Morocco in an age of neoliberal government." *Review of African Political Economy* 38(129): 403–417.

3

NEW AFRICAN CITY PLANS

Local urban form and the escalation of urban inequalities

Vanessa Watson

Introduction

Currently Africa's larger cities are being subjected to new forces and dynamics that are having a profound impact on their socio-spatial growth and change (see Murray – Chapter 2 of this book; Watson 2014). A combination of rapid population increase in these cities as well as economic growth fuelled by the current resource boom has meant an escalation in the demand for urban land, and particularly land that is well located and serviced. A still small but steadily growing middle-class is generating often speculative private sector property development, which in turn has its own logic of urban land demand. One outcome of these new land pressures has been a flurry of new city plans (usually termed 'master plans') frequently developed by private sector architecture and property development companies and portrayed on company websites. While Chapter 2 by Martin Murray in this book describes these new plans in more detail and offers examples of some of them, this chapter will focus on the likely socio-spatial impact of the plans, should they come to fruition.

The chapter will argue that income inequalities are already increasing dramatically in most African countries. The spatial form of city growth (anywhere in the world) is a factor which influences, and is influenced by, income and welfare. Where cities are experiencing rapid growth, and are inevitably subject to major infrastructural and land decisions and investments, then income inequalities will be significantly affected. Moreover, the land and infrastructure decisions which are made now in these cities will set the pattern of socio-spatial inequalities well into the future. Urban income inequality can have serious social and political implications, often giving rise to increased crime, violence and political disruption, and these in turn can hold back economic growth and development. The nature of most of the new plans emerging on the African continent will almost inevitably exacerbate these socio-spatial inequalities and will be very hard to counter.

The chapter will first set the context by describing the nature of the urban transition which is underway in Sub-Saharan Africa, together with the issues which will be confronting cities into the future. It will then examine the question of socio-spatial inequality which has emerged in many of these cities, in large part due to the colonial and post-colonial nature of urban planning and regulation. Finally the chapter will make the argument for how the new 'master plans' are likely to dramatically reinforce and exacerbate the inequalities that are currently emerging. The counter-argument to this is that African cities need to be guided by a very different set of values and priorities if they are to avoid futures that are disastrous in the extreme.

The African urban transition

Sub-Saharan Africa is the last of the world's continents to undergo an urban transition. At the time of colonial independence, most of Africa's population was rural and it is still the least urbanized region in the world. However, Sub-Saharan Africa's urban population is expected to double in the next 20 years, and will be 50 per cent urbanised by 2035 compared to 2020 in the case of Asia with its longer history of urbanization (UNPD 2012). It already has a total urban population larger than that of North America (UNFPA 2007). Africa's urban growth is occurring within a continuum of settlement types including the largest cities, secondary cities and towns. While most attention is still given to the growth of megacities and capital cities, there are indications that in Africa, as in other urbanising parts of the world (UNFPA 2007), the majority of urban growth is occurring in intermediate and smaller cities. Cities with less than 500,000 people currently host 57 per cent of the African urban population (UNPD 2012) and absorb two-thirds of all urban population growth, with this proportion only expected to increase (UN Habitat 2008, 2010). Lagos (Nigeria) is the only city in Sub-Saharan Africa classified as a megacity, and in the category of megacities with more than 10 million people it is one of the smallest. It contained 11.2 million people in 2011 but is growing at a faster rate (3.71 per cent pa) than all of the others (DESA 2012).

A further often-neglected factor is that only a third of this urban growth is a result of rural to urban migration, with the other two-thirds resulting from the natural increase of the population already urbanized (UN Habitat 2010). Hence, over the next 10 years 50 million people will leave the countryside to move to cities, but a further 100 million people will be added to current urban populations from the expansion of families already living in towns and cities. While politicians in Africa and even some international development agencies still sometimes refer to the need to 'curb urbanization it is clear that this is an impossible hope on their part and that the urban transition is being driven forward by factors well beyond the control of policy-makers.

Not only is Africa's urban transition occurring later and more rapidly than it has in other parts of the world, but it has a number of other distinctive features as well. One of these is the precise mode of economic production associated with urban processes (see Gollin et al. 2012, Henderson et al. 2013, Jedwab 2012). In contrast to other regions where cities exist as sites of production, African cities are better understood as 'consumer cities', with urbanization driven by natural resource exports. Those African countries which rely heavily on natural resource production suffer from a 'resource curse' – growth is based on extracting commodities and export proceeds pay for imported consumer goods. This does not generate proportional employment, either within cities or anywhere else. This feature underlies a further important aspect of African cities: where economic growth is not producing formal urban jobs, then the majority of urban populations have little alternative but to survive off informal income-generating activities, and find shelter in informal settlements. UN Habitat (2008) estimated that 62 per cent of urban populations in Sub-Saharan Africa live in 'slums' (compared to a much lower 43 per cent in South Asia and 27 per cent in Latin America) and 70 per cent earn an income through informal activity. A lack of formal jobs impacts particularly heavily on the urban youth with only 17 per cent of young people of working age having access to full-time wage employment in low-income African countries. Given that Africa has the fastest growing 0–14-year age cohort in the world, the impact of jobless economic growth on this youth bulge is of great concern. It implies as well a rapidly growing population which will require access to education, health care services, housing and so on, primarily in the urban areas (Pieterse and Parnell 2014), and a very limited possibility that these will be provided.

The rapidity with which Africa's urban population will double (20 years) has major implications for the supply of basic necessary infrastructure such as water and sanitation. Not only are there currently significant backlogs in many cities, but even if it was possible to catch up with current demand, these capacities would need to double again in a very short space of time (International Bank for Reconstruction and Development 2012). Where growing populations, largely poor, do not have access to clean water and sanitation, then health impacts are unavoidable. The mutually reinforcing relationship between poverty, poor services and poor health is well known (see Mitlin and Satterthwaite 2013).

The section above has served to emphasise the very serious developmental problems faced by African cities and how this challenge will escalate in the decades to come. In cities where the majority of the population is poor, unemployed and unserviced, the hope of many African politicians and policy-makers that these cities can become 'world-class' through the implementation of modernist urban plans and particular building styles (as discussed in Chapter 2 of this book) is clearly not only unrealistic but also a major distraction from the real problems these cities face. The following section moves on to argue that past and current spatial planning approaches in many of these cities have been highly inappropriate and have served to create major spatial inequalities. These inequalities will be worsened considerably should the new fantasy master plans be implemented.

African urban spatial planning: entrenching colonial inequalities

In many parts of the world, including Africa, planning systems are in place which have been imposed or borrowed from elsewhere. In some cases these 'foreign' ideas have not changed significantly since the time they were imported. Planning systems and urban forms are inevitably based on particular assumptions about the time and place for which they were designed, but these assumptions often do not hold in other parts of the world and thus these systems and ideas are often inappropriate (and now often dated) in the context to which they have been transplanted. Frequently, as well, these imported ideas have been drawn on for reasons of political, ethnic or racial domination and exclusion rather than in the interests of good planning.

Colonialism was a very direct vehicle for the spatial translation of planning systems, particularly in those parts of the world under colonial rule when planning was ascendant. In these contexts planning of urban settlements was frequently bound up with the 'modernising and civilising' mission of colonial authorities, but also with the control of urbanization processes and of the urbanising population. On the African continent this diffusion occurred mainly through British, German, French and Portuguese influence, using their home-grown instruments of master planning, zoning, building regulations and the urban models of the time – garden cities, neighbourhood units and Radburn layouts, and later urban modernism. Most colonial and later post-colonial governments also initiated a process of the commodification of land within the liberal tradition of private property rights, with the state maintaining control over the full exercise of these rights, including aspects falling under planning and zoning ordinances.

Significantly, however, imported colonial planning systems were not applied equally to all sectors of the urban population. For example, towns in Cameroon (Njoh 2003) and in other colonised territories in Sub-Saharan Africa (Njoh 1999) were usually zoned into 'low-density residential areas' for Europeans (these areas had privately owned large plots, were well serviced and were subject to Europe-style layouts and building codes); 'medium-density residential areas' for African civil servants (with modest services, some private ownership and the enforcement of building standards); and 'high-density residential areas' (for the indigenous population who were mostly involved in the informal sector), with little public infrastructure, and few or no building controls. Spatially the low-density European areas were set at a distance from the African and Asian areas, ostensibly for health reasons.

Many African countries still have planning legislation based on British or European planning laws from the 1930s or 1940s, which has been revised only marginally. Post-colonial governments tended to reinforce and entrench colonial spatial plans and land management tools, sometimes in even more rigid form than colonial governments (Njoh 2003). Enforcing freehold title for land and doing away with indigenous and communal forms of tenure was a necessary basis for state land management, but also a source of state revenue and often a political tool to reward supporters. Frequently, post-colonial political elites who promoted these tenure reforms were strongly supported in this by former colonial governments, foreign experts and international policy agencies.

Important and capital cities in Africa were often the subject of grand master planning under colonial rule, sometimes involving prominent inter-national planners or architects. Remarkably, in many cases, these plans remain relatively unchanged and some are still in force. The guiding 'vision' in these plans has been that of urban modernism, based on assumptions that it has always been simply a matter of time before African countries 'catch up' economically and culturally with the West, producing cities governed by strong, stable municipalities and occupied by households who are car-owning, formally employed, relatively well off and with urban lifestyles similar to those of European or American urbanites. However, the most obvious problem with these forms of urban modernism is that they fail to accommodate the way of life of the majority of inhabitants in rapidly growing and largely poor and informal cities, and thus directly contribute to social and spatial marginalisation.

Other aspects of urban modernist planning also reinforced spatial and social exclusion, and inequality. Cities planned on the assumption that the majority of residents will own and travel by car become highly unequal. The modernist city is usually spread out due to low built density developments and green buffers or wedges. Low-income households which have usually been displaced to cheaper land on the urban periphery thus find themselves trapped in peripheral settlements or having to pay high transport costs if they want to travel to public facilities or economic opportunities. The separation of land uses into zoned monofunctional areas further generates large volumes of movement (as people must move from one to the other to meet daily needs), and, if residential zoning is enforced, this leads to major economic disadvantage for poorer people who commonly use their dwelling as an economic unit as well. The modernist city provides few good trading places for informal sector operators, which today make up the bulk of economic actors in African cities. Street traders are usually viewed as undesirable in the planned parts of cities, particularly around shopping malls, and are strictly controlled or excluded. Yet these are the most profitable locations for street traders as they offer access to the purchasing power of a higher-income market.

The land use regulations which accompany master plans usually demand standards of construction and forms of land use which are unachievable and inappropriate for the poor in cities, which make up the bulk of urban populations in African cities. Hence, high levels of illegality (of buildings and land use) in many cities are a direct result of inappropriate planning and zoning standards, not of criminally minded citizens. Such standards are directly responsible for spatial and social marginalisation. These zoning ordinances, which require high standards, are difficult to enforce where governments lack capacity, often leaving people with no controls or protection at all. In fact there is a growing literature which suggests that the negative impacts of urban modernism and planning regulations on the poor are not just a result of wrongly appropriated, misunderstood or inefficient planning systems, but may in some cases be due to corrupt manipulation or use of the system for political domination. In these cases, it is of course abuse of the planning system rather than the planning system itself which is at fault.

As Cain (2014) has suggested, the resurgence of new fantasy master plans in Africa in the last few years (Chapter 2) echoes the efforts from some decades ago, particularly in the post-independence era. Many governments of the time initiated grand-scale, new capital city building (e.g. Abuja in Nigeria, Dodoma in Tanzania), often for symbolic reasons relating to national pride and a break with their colonial past. These cities drew on the most up-to-date ideas of the time to inform planning, urban design and architecture, and the professionals involved were almost without exception from economically advanced nations. However, 'the new cities often turned out to be expensive mistakes that diverted investment from potentially more sustainable economic and social projects. Moreover, these projects left some of the countries that initiated them with a burden of debt that inhibited national development for a generation or more' (Cain 2014, 2). So while lessons from Africa's past efforts at city building (driven by political ambitions rather than concerns for the welfare of the majority of urban dwellers) are plainly evident, this history has been conveniently forgotten in the new race for 'world-class city' status. The final section of the chapter turns to these new plans and their likely impact.

African urban fantasies in the current era: repeating past mistakes?

This section of the chapter turns to the new African fantasy plans (Chapter 2 of this book and Watson 2014) and considers their likely impact on the problems and issues of African cities described in the first section of the chapter. The new plans have the following in common: they are large-scale in that they involve the re-planning of all or large parts of an existing city, or (more often) restructuring a city through the creation of linked but new satellite cities; they consist of graphically represented and three-dimensional visions of future cities rather than detailed land use plans and most of these visions are clearly influenced by cities such as Dubai, Shanghai or Singapore; there are clear attempts to link these physical visions to contemporary rhetoric on urban sustainability, risk and new technologies, underpinned by the ideal that through these cities Africa can be 'modernised'; they are either on the websites of the global companies that have developed them or are on government websites with references to their origins within private sector companies; their location in the legal or governance structures of a country is not clear – where formal city plans exist, these visions may simply parallel or over-ride them; and there is no reference to any kind of participation or democratic debate which has taken place.

While detailed research needs to be carried out, the fact that the private sector (with bases in, or links to, economically stronger regions of the world) has become a dominant player in nearly all of these projects (excluding the Chinese-built ghost cities) suggests that global economic forces are interacting with local African contexts in new ways. It is possible to speculate that the downturn in demand for property and urban development in Global North regions after the 2008 financial crisis has driven both built environment professionals and property investment companies to seek new markets in those parts of the world where economic growth and demand for new urban growth continued: particularly in the Middle-East, Asia and Africa. Other local factors that are probably playing a role in the appearance of these new projects are high rates of urbanization and a growing urban middle-class. A number of international finance houses have been highlighting this middle-class growth in African cities. Deloitte[1] states that Africa's middle-class has tripled over the last 30 years and is now the fastest growing in the world. A growing urban middle-class generates demand for formal housing, public facilities and amenities, retail outlets and transport routes for growing car-ownership, and certainly potential customers for the kinds of urban environments portrayed in the urban fantasy plans referred to above. This class also provides a consumer market for goods and services of all kinds and hence investment in production and services buildings. A growing middle-class therefore fuels demand for well-located and serviced urban land and development projects as well as architectural styles considered 'aspirational' or 'modern'. But Deloitte is cautious on the definition of a middle-class. They note that the African Development Bank uses as a definition those spending between US$2 and 20 a day, and even the 'upper middle-class' as US$10 to 20 a day spenders. It is difficult to imagine how households with such minimal spending power can afford the luxury apartments portrayed in the fantasy plans (as well as the vehicles needed to move around them) and (as in the case of Luanda's Chinese ghost cities) it may be that prospective property developers are seriously misreading the African market.

In looking at the possible impact of these new plans on Africa's cities, and the likelihood that they may address the key issues of African urban growth, it is possible to consider other parts of the world where plans and urban images of a similar kind have been in circulation much longer than they have in Africa (see Bhan 2014). For example, Goldman (2011) explains how in Bangalore newly created parastatals designed to fast-track particularly large projects are externally funded and have little or no local oversight, and hence local government has been carved up into the older bureaucracies left in charge of small maintenance budgets and the new autonomous agencies fed by international loans but also large obligations of risky debt finance. Projects being dealt with in this way are the new Bangalore airport, an 'IT corridor' and overhead freeways to connect them, as well as the familiar satellite cities: for example, Knowledge City, built on working farmland, is financed by a Dubai firm. Here 200,000 rural people were displaced by the Mysore-Bangalore project with minimal compensation for what was called unproductive farmland, but this was land that was immediately turned into high-value urban land. Many dispossessed land-owners, especially women, found it very difficult to prove their ownership, or were tenants with no claim to the land. Mass removals of populations on or beyond the urban periphery inevitably swelled the ranks of those already in urban informal settlements.

In Africa these processes are just beginning. The impact on poorer urban dwellers will be felt most directly where new urban master plans and projects attempt comprehensive urban renewal to remake the city in the image of somewhere else considered 'world-class'. Kigali and to a lesser extent Addis Ababa seem to be currently subject to these kinds of make-overs, and their extensive shack populations are being systematically moved to make way for the new projects. Many other cities (such as Nairobi) are responding to the very real problem of traffic congestion by planning new systems of freeways and fly-overs which carve their way through older and poorer urban areas. These cater directly for the still small car-owning middle-class, but are of little help to the majority of people who travel on foot.

In most cities, however, governments find it easier to avoid the difficulties of removal of dense urban fabric and to seek less fiercely contested land on the urban edge (Lagos, Dar es Salaam, Kinshasa) or in the rural areas beyond. Around African cities, peri-urban areas have been growing very rapidly as poor urban dwellers look for a foothold in the cities and towns where land is more easily available, where they can escape the costs and threats of urban land regulations, and where there is a possibility of combining urban and rural livelihoods. These are the areas usually earmarked for development by new urban extension projects.

Writing about the Cité le Fleuve project in Kinshasa, De Boeck (2012) describes how colonial and post-colonial planned expansions of Kinshasa have over the years been re-territorialised and reclaimed by poorer city inhabitants, redefining the colonial logics that were stamped onto this space. Large tracts of land along the Congo River have been converted into productive rice fields which supply Kinshasa's markets, although more recently pressures of urban growth have seen some of these areas converted to shacklands. Acknowledging that this existence outside of the official frameworks of formal urban regulation and services is not an ideal way to live, De Boeck (2012) nonetheless argues that it allows the pursuit of livelihoods with a degree of freedom and flexibility. In other terminology this would be called 'resilience'. All this stands in contrast to the recent initiatives to 'modernise' Kinshasa, starting with the conversion of tree-lined boulevards to an 8-lane highway into the heart of the city, and efforts to 'sanitise' the city by expelling street children and small traders (a 'politics of erasure'). Cité le Fleuve on reclaimed land in the Congo River is a continuation of this modernising effort, but in the process will destroy much of the rice-producing areas and the economic networks they support.

In the case of satellite cities, these are frequently justified as being located on 'empty land', but it is rare that land around larger cities is empty, and if such land is not within an environmentally protected area then it is very likely to be actively farmed. In all these kinds of eviction processes land-owners rarely hold land title, and full compensation for land, shelter and livelihoods is rare. For land-owners near Nairobi's satellite Konza City, Baraka Mwau (2013) writes about their anticipation that the new city may create construction jobs for unemployed youth and a demand for their food products. But at the same time their properties fall within the 10-kilometre radius of Konza City and hence are marked for (or not for) demolition, and all new developments need planning approval. In the small town of Malili, Mwau describes a mushrooming of activities and land speculation, with 'plots changing ownership within hours'. He reports that authorities are now doing their best to contain this informality which has sprung up in at least 10 towns near the proposed Konza site. A two-kilometre buffer (cordon sanitaire?) around the City is designated for 'wildlife' and development for eight kilometres beyond this will be controlled. Quite what happens to the land and livelihoods of farmers in this area is not clear.

Beyond these immediate impacts of the new urban developments there are a number of further outcomes which can be anticipated with a degree of confidence. State spending on large-scale infrastructure (transport, sanitation, power) is likely to be skewed in the direction of support for these new cities and projects and away from meeting the basic service and housing needs of the much larger poor urban populations. Should the middle-classes and higher-end investors retreat to these new elite enclaves (and this after all is their target market) then their tax base and spending power will be lost to the city, thus exacerbating urban decline. A very good example of this is the new satellite Kilamba City (a Chinese-built 'ghost town' outside Luanda, Angola). Here new apartments cost between $120,000 and $200,000 – unaffordable even for senior civil servants. Eventually the state was obliged to lower apartment costs through major subsidies which used up the bulk of the national housing budget (Cain 2014), which might have gone to assist poor households in Luanda.

The spatial separation of rich and poor which these new urban fantasies will entrench opens up the prospect of urban spatial and social inequalities at an unprecedented scale. At the same time, the hope that these new cities and developments will be 'self-contained' and able to insulate themselves from the 'disorder' and 'chaos' of the existing cities is remote. Satellite cities are frequently unable to sustain all the job and service requirements of their populations and tend to generate large volumes of movement and traffic as their residents find themselves having to travel back to more established centres. Wealthy enclaves are also usually unable to function without low-income service providers (domestic workers, gardeners, construction workers etc.) and inevitably an informal city grows up around the edges of the formal city.

It is also possible to ask: where are the street traders, the informal transport operators and the shack-dwellers in these new fantasy plans? Would traders ever be allowed to set up business on these pristine boulevards? Could the poor (or even the middle-class) ever afford the glass box apartments, or even have the kinds of jobs that would get them access to the towering office blocks? How do people move around on foot, as most city-dwellers do, through these wide open spaces and car-oriented movement routes? The answer is clearly that all such urban citizens have been swept out of sight in these grand visions. As there is no possibility that the poor majority of African cities will have suddenly become wealthy and skilled, it can only be assumed that they have been chased out of town, removed to land on the city edge and out of sight, or perhaps – as the politicians so often wish – they have gone to smaller towns or rural areas. Where these new visions are satellite cities way outside the main metropolitan area, it is of course the wealthy that will be fleeing beyond the city edge, away from what is perceived as the crime, grime and chaos of the major cities.

In a range of ways the utopian dreams of these urban fantasies (most of which are based on concepts which have been attempted before in other parts of the world and their impacts well documented) are unlikely to materialise, yet the efforts to achieve them will have profound effects on lives and livelihoods. While those with a degree of power and resources may well be able to benefit in various ways, given the overwhelming dominance in African cities of those with very little, a widening and deepening of inequality is inevitable.

63

Conclusions

This chapter has considered the likely impact of the new generation of African urban fantasy plans discussed in Chapter 2 of this book and Watson (2014). Given the almost insurmountable challenges faced by African cities now and in the coming decades (the first section of this chapter), the chapter asks whether these new plans offer any hope that these issues will be addressed. The argument here is that not only do they fail to take account of these, but that they will also directly worsen these urban problems. In doing so, these plans also fail to take account of past attempts to implement rather similar grand urban visions in African cities, where the negative impacts are readily evident. Instead, Africa's larger cities seem to be entering a new era of change, driven by the continent's own economic growth and emerging middle-class as well as an international property development and finance sector in search of new markets. The urban visions and plans which this confluence of interests has produced stand in dramatic contrast to the lived reality of most urbanites, and while their impacts are likely to be complex and contradictory, what seems most likely is that the majority of urban populations will find themselves further disadvantaged and marginalised. It is access to land by the urban poor (as well as those on the urban periphery and beyond) which is most directly threatened by all these processes, and access to land in turn determines access to urban services, livelihoods and citizenship.

Moreover, attempts to implement these new projects are likely to directly divert state infrastructure funds away from dealing with the issue of the doubling of the population in the next 20 years. It is clear that the majority of these coming populations will continue to be young, poor and unemployed, and increasingly politically discontented. Living in poverty alongside the glitzy Dubai architectures of the new projects, it becomes more and more likely that these disaffected populations will publicly and politically express their discontent, making the realisation of 'world-class cities' an even more unattainable dream.

Note

1 http://www2.deloitte.com/content/dam/Deloitte/au/Documents/international-specialist/deloitte-au-aas-rise-african-middle-class-12.pdf

References

Bhan, G. 2014. "The real lives of urban fantasies." *Environment and Urbanization* 26(1): 232–235.

Cain, A. 2014. "African urban fantasies: past lessons and emerging realties." *Environment and Urbanization* 26(2): 561–567.

De Boeck, F. 2012. "Spectral Kinshasha: building the city through an architecture of words." In *Urban Theory Beyond the West: A World of Cities*, by T. Edensor and M. Jayne (eds.), 311–328. London, New York: Routledge.

DESA (Department of Social and Economic Affairs) United Nations Population Division. 2012. *World Urbanization Prospects: The 2011 Revision*. New York: United Nations.

Goldman, M. 2011. "Speculative urbanism and the making of the next world city." *International Journal of Urban and Regional Research* 35(3): 555–581.

Gollin, D., Jedwab, R. and Vollrath, D. 2013. "Urbanization with or without structural transformation." The National Bureau of Economic Research (NBER) Growth Conference. San Francisco: NBER.

Henderson, J.V., Roberts, M. and Storeygard, A. 2013. "Is urbanization in Sub-Saharan Africa different?" Policy Research Working Paper 6481 (The World Bank) 1–46.

International Bank for Reconstruction and Development. 2012. *The Future of Water in African Cities: Why Waste Water?* Washington D.C.: The World Bank.

Jedwab, R. 2011. "Why is African urbanization different? Evidence from resource exports in Ghana and Ivory Coast." Job Market Paper (Paris School of Economics and STICERD, London School of Economics) 1–60.

Mitlin, D. and Satterthwaite, D. 2013. *Urban Poverty in the Global South: Scale and Nature*. London: Routledge.

Mwau, B. 2013. "Blog: the planned hatches the unplanned." 2 August. Accessed on 28 July 2016 from http://slumurbanism.wordpress.com/2013/08/02/the-planned-hatches-the-unplanned.

Njoh, A. 1999. *Urban Planning, Housing and Spatial Structures in Sub-Saharan Africa*. Aldershot: Ashgate.

——. 2003. *Planning in Contemporary Africa: The State, Town Planning and Society in Cameroon*. Aldershot: Ashgate.

Pieterse, E. and Parnell, S. 2014. "Africa's urban revolution in context." In *Africa's Urban Revolution*, by E. Pieterse and S. Parnell (eds.), 1–17. London: Zed Books.

UNFPA (United Nations Population Fund). 2007. *State of World Population 2007: Unleashing the Potential of Urban Growth*. New York: United Nations.

UN Habitat. 2008. *The State of African Cities 2008: A Framework for Addressing Urban Challenges in Africa*. Nairobi: United Nations.

——. 2010. *The State of African Cities 2010: Governance, Inequality and Urban Land Markets*. Nairobi: United Nations.

UNPD (United Nations Department of Economic and Social Affairs Population Division). 2012. *World Urbanization Prospects: The 2012 Revision*. New York: UNPD.

Watson, V. 2014. "African urban fantasies: dreams or nightmares." *Environment and Urbanization* 26(1): 215–231.

4

SPEED KILLS

Fast urbanism and endangered sustainability in the Masdar City project

Federico Cugurullo

Introduction

There is much empirical evidence that the planet is increasingly becoming an urban construct. Over the last couple of decades, the scope and speed of urbanization have increased exponentially, pushing the boundaries of what scholars such as Merrifield (2012) define as *planetary urbanization* (see also Urban Age Project 2007, 2011). This phenomenon can be understood as a relentless growth of urban fabric across the world: fabric that covers and underpins the surface of the Earth, reshaping nature–society relations with substantial environmental, economic and social impacts.

Far from being a homogeneous phenomenon, the urbanization of the planet is characterized by an amalgam of divergences and convergences that form a complex urban mosaic made of a plethora of heterogeneous tiles. In the Global South, for instance, the production of built environments takes a multitude of shapes, such as new master-planned cities, retrofits, urban expansions and regenerations, which manifest considerable differences in terms of form and substance. From a political perspective, these diverse urban shapes are the product of different institutional settings, ranging from authoritarian regimes to democratic and post-democratic institutions, which fund, plan and govern new urban spaces on the basis of heterogeneous ideologies. From an architectural point of view, the multiple forms of the urbanization of the Global South are shaped by the imaginations and dreams of local and international architects, and are influenced by contextual challenges and practicalities such as climate, political stability and economic resources.

However, within these variegated processes of urbanization it is possible to find strategies of city-making driven by similar ideals of planning, design and architecture, and trace the contours of homogenous urban trends. The creation of new master-planned cities, for example, appears to be an emerging phenomenon in the Global South: a phenomenon that is often inspired by a single, recurring concept: the *eco-city*. Over the last twenty years, across developing countries, nineteen projects for new eco-cities have been conceptualized and developed and, today, their implementation represents one of the key faces of the urbanizing South of the world (Cugurullo 2013).

In the Global South, the genesis and development of eco-city projects manifest fast dynamics. Speed is one of the key factors that characterize projects such as Masdar City, Tianjin, Caofeidian and Songdo. These cities are not the product of gradual socio-political and urban transformations. Instead, they originate from fast processes of capital accumulation, urban planning and city-building. Eco-city projects emerge as fast solutions to pressing economic, environmental and political challenges, ranging from resource scarcity and climate change to institutional instability, and, as such, they become 'fast' cities: settlements conceptualized in a couple of years and built from scratch over a couple of decades.

This chapter argues that, specifically in relation to urban sustainability, speed is a problem. Planning and architectural velocity, understood as the rapid production and assembly of urban fabric, hinders the social and environmental development of fast cities. Projects for new eco-cities manage to reach and keep high speeds by focusing on specific aspects of urbanization tailored around pre-determined economic and political goals. An eco-city project like Masdar City, for example, focuses on the development of urban technologies to regenerate the economy of Abu Dhabi and create alternative energy sources. However, by investing the majority of their financial power in the production of urban technology, developers neglect important ecological and social issues, such as the preservation of regional ecosystems and the cultivation of urban communities. In essence, the key argument and message of this chapter is that *speed comes at a price*. Urban history shows how urban development takes more than a few decades. The formulation of efficient strategies of urban–economic growth, the conservation of natural environments and the formation of cohesive societies are complex socio-environmental and political processes that require many years of contest, struggle and negotiation. The quick fix advanced by eco-city projects leads stakeholders to implement superficial and partial urban policies that fail to equally target economies, societies and natural environments to the detriment of the sustainability of the new cities.

The following sections explore the development of Masdar City, a new eco-city project in Abu Dhabi, to shed empirical light on eco-city initiatives in the Global South. The analysis focuses on two interconnected themes which form the main sections of the chapter: the rationale of an eco-city project and the fast urbanism that supports its development. After providing an overview of the eco-city phenomenon in the developing world and introducing the key features of Masdar City, the chapter explores the economic mechanics of the project. It examines the clean-tech business that is behind the genesis and development of the new city and explains how the Emirati eco-city initiative generates economic returns by partnering with leading companies in clean energy and technology. The analysis then shifts to the planning strategies that underpin the implementation of Masdar City, stressing how the physical development of the new city is dictated by the economic imperatives set by stakeholders whose interests control the evolution of the master plan. In the second half of the chapter, the discussion focuses on the speed of the Masdar City project. The concept of speed is approached from architectural, planning and regulatory perspectives and frames an exploration of the implementation of the Emirati eco-city project, with an emphasis on the limits of fast urbanism. Finally, Masdar City is understood as the product of economic rationales that hinder the environmental and social dimensions of the city, thereby penalizing its overall performance and sustainability potential.

The analysis draws upon empirical research carried out in Abu Dhabi from September 2010 to May 2011. The data discussed in the following sections was produced through the combination of twenty in-depth semi-structured interviews with key actors involved in the Masdar City project (developers, business partners, leading architects and planners) and a critical documentary analysis of material concerning the design and planning strategies of the new city (master plans and construction reports). For privacy purposes, anonymity was granted to all the interviewees and, in the text, the names of the participants have been substituted by their role in the project.

New eco-city projects in the Global South

There is no established definition of *eco-city*. As noted by Rapoport (2014), the few attempts to define an eco-city tend to be generic and limited to lists of principles. The term 'eco-city' was coined by Richard Register (1987, 2006) in the late 1980s and refers to peculiar ideas of space and society–nature relations. Register, an environmental activist and theorist, wanted to break the historical divide between the built and natural environment, and advanced a vision of the city as a compact space sensitive to local ecologies. However, as critical urban studies have shown, there is a broad line that separates what an eco-city is in theory and what eco-city projects are in practice. Chinese and Emirati eco-city projects such as Dongtan, Chongming and Masdar City, for instance, have been critiqued as mere tools of capital accumulation, largely insensitive to important environmental questions (Chang and Shepherd 2013, Cugurullo 2013a, 2013b). Similar concerns have been expressed in relation to Indian and South Korean eco-city initiatives like Lavasa and Sondgo where economic objectives appear to have been prioritized over ecological targets (Datta 2012, Shwayri 2013).

At the time of writing, nineteen projects for new eco-cities are under development across the Global South.[1] These projects appear to be characterized by a series of convergences and divergences. On the one hand, eco-city initiatives seem to be inspired by similar ideas of urban planning and sustainability. The ideal of sustainability advanced by eco-city developers reflects an understanding of sustainable development grounded in three interconnected elements: environment, economy and society. Drawing upon the logics of the Triple Bottom Line (see Elkington 2004) and the Brundtland Report (WCED 1987), eco-city developers advocate a balance among economic, social and environmental concerns, and depict the built environment as the medium through which to realize an equilibrium between society and nature. The pursuit of this urban ideal is carried out via master planning. New spaces are envisioned, designed and implemented. Cities are built from scratch, and what previously were underdeveloped areas are transformed into hyperactive centres of human activity and capital accumulation.

Technological innovation plays a key role in eco-city formations. New technologies provide the physical tools by means of which urban sustainability is intended to be achieved. Cutting-edge devices such as smart-grids are extensively employed by eco-city developers with the aim of reducing urban carbon emissions. As Joss and Molella (2013) note, new eco-city projects like Caofedian (China) can be defined and understood as *techno-cities*: 'cities planned and developed in conjunction with large technological projects' in which the production of urban space follows and is followed by the production of technology (Kargon and Mollela 2008, 11). Echoing the thesis of ecological modernization, projects for new eco-cities profess a synergy between economic development and environmental conservation, and see in technological advancement the opportunity to boost local economies without damaging the environment.

On the other hand, such ideological similarities seem to fade once eco-city projects are implemented, and planning theory gives way to practice. Place is central in shaping the implementation of projects for new eco-cities. First, alleged eco-cities are built *ad hoc* on the basis of particular economic, political and environmental challenges. Projects for new eco-cities are rarely isolated urban experiments. Instead, they can be interpreted as part of broader policy agendas designed by policy makers to meet regional development targets (Cugurullo 2015). Second, eco-city initiatives are directly exposed to the context that surrounds their genesis and development and, therefore, they bear the mark of specific cultures, economies, political institutions and biophysical environments. As a result, across the Global South there is not one, but a variety of different and often colliding examples of eco-city formations whose divergences become evident when the projects are observed through the lens of geography. The construction of King Abdullah City for Atomic and Renewable Energy (K.A.CARE), for example, manifests the recent interest of Saudi Arabia in atomic energy. The new Saudi city finds its genesis in increasing pressures on the country's energy resources, due to high levels of population growth.

In this sense, K.A.CARE represents the attempt of Saudi policy makers to develop alternative energy sources, and emerges as a unique architectural and engineering blend of atomic, photovoltaic, wind-power and geothermal technologies.

The chapter now turns to an in-depth examination of a single project for a new eco-city, Masdar City, and uses the Emirati example to explore the economics and planning of an eco-city project to show how, in Abu Dhabi, the Masdarian urbanism sacrifices sustainability for the sake of speed. The narrative underpinning the following sections does not enter the sustainability debate and restrains the analysis from discussing what sustainability is or is meant to be. Specifically in relation to urban planning and politics, the paradoxes of sustainability have been extensively discussed by a number of scholars (see, for instance, Brand and Thomas 2005, Krueger and Gibbs 2007). In relation to eco-city projects, it has been empirically proven that sustainability has emerged mostly as a label: an empty label employed by stakeholders to cover ecologically unfriendly economic objectives and business interests (Cugurullo 2013a, 2013b). Yet the concept of sustainability, in its original scientific and philosophical sense, broadly intended as the capacity to *sustinere* (sustain) equal society–nature relations, can still be a precious analytical tool to shed light upon the socio-environmental and economic performances of alleged eco-cities, and this is how the term will be employed in the remainder of the chapter.

The Masdar City project

Masdar City is an ambitious project for a new master-planned eco-city currently under development in Abu Dhabi: the leading emirate of the United Arab Emirates (UAE). The project began in 2006 when Abu Dhabi announced the genesis of what was alleged to become the first zero-carbon city in the world and a global paradigm of urban sustainability. Foster and Partners realized the first master plan whose implementation started in 2008 under the aegis of the Masdar Initiative: a state-owned company formed by the government of Abu Dhabi to promote the research and development of clean energy solutions and technologies. From an architectural perspective, the new city adopts an iconic design based on a perfect square. The perimeter is formed by four walls intended to provide cover from the desert and separate the settlement, physically and symbolically, from the rest of Abu Dhabi. Spatially, Masdar City covers an area of 6 km^2 and is located close to Abu Dhabi International Airport, approximately 15 km from the capital city of Abu Dhabi. Once completed, the settlement is intended to have a population of 40,000 residents, and be able to supply itself by means of a vast array of clean technologies (such as concentrated solar power and geothermal stations) that, at the time of writing, are being extensively integrated in the physical infrastructure of the new city.

Masdar City is not an isolated urban project. The construction of an eco-city is part of a broader Emirati policy agenda seeking to stimulate economic growth through urban growth. The development of Abu Dhabi's eco-city project is inscribed into a particular spatio-temporal context shaped by strong economic and political pressures. With its oil reserves close to depletion and the specters of the Arab Spring looming on the Islamic world, today Abu Dhabi is seeking to preserve its political and economic foundations via urbanization. As emphasized in recent official publications, the government of Abu Dhabi (2008) is aware of the instability of the economy of the region and has expressed the will to develop alternative energy sources and economic sectors. According to governmental projections, the Emirati oil reserves will be capable of sustaining the local economy until the end of the twenty-first century but, by then, the government aims to have already established a post-petroleum economic system based on clean, renewable energies (Cugurullo 2015). In this context, the Masdar City project wants to be a double power source. From an economic perspective, the new city seeks to attract companies and capital to Abu Dhabi by developing a new strand of the economy grounded in the production of clean technologies. From an energy perspective, Masdar City aspires to be a large-scale power plant generating vast amounts of clean energy via urban technology.

The rationale of Masdar City

The nucleus of the Emirati eco-city project is made of interconnected economic mechanisms whose examination is key to understanding how and why the concept of the eco-city has become part of the urban agenda of Abu Dhabi. Masdar City is a platform: a vast urban platform in which new technologies can be researched, developed, tested and showcased. The focus of the developers is on clean technology: cutting-edge photovoltaics such as concentrated solar power, geothermal stations, smart-grids, smart appliances and automated transport systems, that the Masdar Initiative, in partnership with established and emerging clean-tech companies, produces and commercializes for economic purposes.

The Masdar City project is based on three interconnected elements. First, the *built environment* which provides the infrastructures necessary to develop, test and showcase clean-tech products. Laboratories, offices and showrooms make up the majority of the urban fabric of the new city. Residential areas are a minor part of the project and 80 per cent of them are assigned to the people that work in Masdar City as either members of the Masdar Initiative or business partners. Second, *expertise*, which the Masdar Initiative provides in the shape of engineers, scientists, IT experts, market analysts and academics specialized in clean technology, renewable energy, economics of carbon credits and sustainability issues. These experts are

71

assigned to the clean-tech projects carried out by the Masdar Initiative and its partners, and work on the development and circulation of the Masdarian technology. Third, *funding schemes* made available via Masdar Capital: a special unit of the Masdar Initiative 'created to support the development of new technologies and projects and generate positive returns for Abu Dhabi' (2014, no page). Through these schemes, the Masdar Initiative supports the development of some of the clean-tech products that are created and tested in Masdar City. Average investments range from $15 to 50 million and tend to target emerging companies in the clean-technology economy.

Each technology developed in Masdar City is linked to a clean-tech project that, in turn, is linked to a business partnership between the Masdar Initiative and a clean-tech company. In the Masdar City project, business partnerships lead to two main economic outcomes. First, the production and commercialization of clean technologies generate revenues that the Masdar Initiative shares with its business partners. Every partnership is associated with the development of a specific technology. Mitsubishi, for example, is working with the Masdar Initiative on low-carbon electric vehicles, while companies like Siemens and Schneider are in Masdar City to develop and test smart power grids designed to synchronize energy demand and supply. When these technologies are sold, the Masdar Initiative shares the profit with the company with whom the product has been developed. 'Mass-production' is the expression used by a manager of the Masdar Initiative, during an interview, to describe the target of the Emirati eco-city project. The objective of the Masdar City project is to develop and commercialize clean-tech products that will sell worldwide: an objective that is supported by a vast promotional campaign including bombastic public events such as the World Future Energy Forum, and a large web portal called The Future Build where the Masdarian technology can be purchased.

Second, partnerships generate rents. The Masdar City project is entirely funded by the government of Abu Dhabi which is the only owner of the new city. If a company wants to be based in Masdar City, it has to pay for it. More specifically, companies such as Siemens, Schneider and Mitsubishi can rent a portion of the total area occupied by Masdar City and use it to build their offices, laboratories and showrooms. What is built remains the property of Abu Dhabi that acts as a landlord. Renting costs vary according to the dimension and location of the buildings, the materials employed in their construction and the type of architecture and design that companies want to implement. 'Here we are simply tenants,' a representative from Schneider explained in an interview. According to a manager from the Masdar Initiative, to date Masdar City is the biggest urban development in Abu Dhabi and, once its construction is completed, the profit from its lease is expected to overcome that from any other property in the region.

The implementation of Masdar City

The new Emirati city is designed to evolve according to the business partnerships forged by the Masdar Initiative, and the technologies that these partnerships, in turn, generate. In the Masdar City project, the portfolio of partnerships varies through time, and so does the master plan. The Masdar Initiative is constantly looking for new business partners and it expands the new city only when new partnerships are secured. To paraphrase the comment of a planner from Foster and Partners, Masdar City is built *on demand*. Unless there are clean-tech companies ready to move to Masdar City, the planning stage enters a condition of stasis: construction stops, workers leave the site and the part of the city that is under development is forced into a state of unconsciousness. This artificial coma stops the moment a company signs a partnership with the Masdar Initiative. Only then is the planning machine set in motion again, and a new area of the city is designed and then built specifically to meet the expectations of the new business partner.

The planning of Masdar City is directly influenced by the economics of Masdar City. In 2007, Foster and Partners designed the master plan of the new Emirati city, drawing upon traditional Islamic architecture and urban planning. The idea of the studio was to create a pedestrian city composed of narrow streets meant to channel the wind, provide shade and reduce the perceived temperature. The original master plan featured a car-free city in which streets were designed exclusively for pedestrian purposes; a public transport system, composed of automated electric vehicles, was meant to operate underground.

However, the original version of the master plan will never be realized for two key reasons. First, in the beginning of the project, the Masdar Initiative decided to keep the master plan of the new city flexible to accommodate the interests of future business partners. As explained earlier in this section, the clean-tech companies that join the Masdar City project have the possibility to claim part of the city and rent it. In addition, the business partners of the Masdar Initiative can choose the size and design of their buildings according to their imagination. For this reason, every time a lease is signed, the master plan has to change and adapt to the will of the new business partner. The objective of the Masdar Initiative is to make the new city as appealing as possible to keep attracting new companies. A spokesman of the Masdar Initiative explained in an interview that, today, the majority of multinationals have specific ideas of space: ideas that they want to see implemented in all their buildings. Masdar City seeks to support this trend by offering its partners in business the possibility to maintain their philosophy of design. For companies like Siemens and Schneider, this was one of the main factors that influenced their decision to become part of the Masdarian venture. A member of Siemens stated in an interview that having a distinct architectural style is a top priority for his company, and admitted that the liberty granted by the Masdar Initiative in terms of design and architecture was a unique selling point.

Second, Masdar City evolves following the types of technology that stakeholders decide to develop. The new Emirati city is intended to be a living urban laboratory to test clean technologies. Therefore, the physical characteristics of Masdar City have to suit the physical characteristics of the clean-tech products that will be tested. An emblematic example is that of Mitsubishi which signed a partnership with the Masdar Initiative in 2011 to research and develop electric cars. From a planning perspective, cars, regardless of what energy supplies them, require particular urban spaces and infrastructures that often differ from those used by pedestrians. Until 2011, Masdar City was still supposed to be a car-free city and the master plan did not feature any street accessible for cars. Therefore, when Mitsubishi joined the project, Masdar City needed to change. Foster and Partners had to radically modify their vision of the city, and include spaces where cars could be driven and tested.

The remainder of the chapter looks at Masdar City from the perspective of speed. As explained earlier, Abu Dhabi is keen to implement the Masdar City project as soon as possible and secure alternative sources of capital and energy. By looking at the processes of planning and regulation that underpin Masdar City, the chapter now examines how the Emirati eco-city project has been quickly conceptualized and adjusted to the local building code, and reflects on the negative socio-environmental externalities of the Masdarian fast urbanism.

'Fast' planning

The planning process of Masdar City is characterized by fast dynamics. The formulation of the design, architecture and physical arrangement of the new Emirati city has advanced at a fast pace from the start. In early 2007, when the project for a new settlement was approved by the government of Abu Dhabi, the Masdar Initiative quickly launched a design competition for a new sustainable city. The competition was won by Foster and Partners that proposed a revolutionary, car-free urban space designed for a pedestrian population. As the chief-planner from Foster and Partners recalled in an interview, 'the process was fast'. The Masdar Initiative did not advance any specific idea of urbanism and architecture, nor did it make any request. The planner explained that the developers simply asked for an 'eco-city', giving to the architectural studio complete intellectual freedom. Foster and Partners worked independently to finish the first master plan of what was soon to be named 'Masdar City'. According to the chief-planner, the role of the Masdar Initiative was very limited. The Emirati company wanted to begin the construction phase 'as soon as possible' and did not delay the work of the studio. The planner claimed that when the first draft of the master plan was presented, the Masdar Initiative made no corrections and the project quickly moved to the next stage: implementation.

Once the Masdar City project entered the construction phase, the attitude of the developers remained the same. The Masdar Initiative set a two-year deadline for completing the first part of the project (the Masdar Institute of Science and Technology), and became less and less interested in the conceptual aspects of the development, such as the meaning of sustainability in Masdar City and how the new city was going to tackle environmental issues from an architectural and planning perspective. Interview data with representatives from Foster and Partners and the Masdar Initiative showed that the latter did not have a specific 'eco-city' ideal. Instead, what the Masdar Initiative advanced was a labile urban ideal. The developers aimed to keep their vision of sustainability flexible to adapt it to potential changes within the business portfolio of their partners. Following this philosophy, the team of architects and planners from Foster and Partners had to keep the master plan flexible, and instead of crystallizing a precise and rigid plan, they produced a loose and adaptable, long-term planning strategy made of different, potential incarnations of Masdar City.

From a planning perspective, Masdar City exists and does not exist at the same time. The planners behind the project have a series of planning and architectural options and, eventually, the solution that is selected is the solution that works best for the Masdarian business. What is important to emphasize is that there is no ultimate sustainability goal or ideal according to which the urban fabric of the new settlement is shaped. It is the economic dimension of Masdar City that forms the core of the project and defines every other aspect of the city. Business interests and economic agendas shape the physical incarnation of the Emirati eco-city project, dictating its architecture and design. However, because business plans are as mutable as capital, the layout of Masdar City is also unstable and unpredictable, and it is impossible to foresee what final shape the new city will eventually take. It is emblematic that, during an interview, the chief-planner from Foster and Partners was unable to describe how Masdar City was going to look and be like in thirty years. Neither Foster and Partners nor the Masdar Initiative can visualize the urban future of the Masdar City project. What is clear to the developers is the economic target of the project, and it is towards that target that the Masdarian eco-city initiative is moving fast.

Modular planning

Masdar City has been conceptualized to be operative from the early stages of the project. As explained in the previous section, to prevent a potential political collapse, Abu Dhabi needs to develop a new economic sector as well as new sources of energy as soon as possible. However, from a planning perspective, the construction of a new city is a long process and even the implementation of a small settlement such as Masdar City might take more than a couple of decades. Abu Dhabi cannot wait that long to regenerate its economy and energy sources. For the local elites, the development of the eco-city project has to be fast. For this reason, Masdar City has been designed according to a modular rationale. The master plan is divided into different phases that correspond to different, independent areas of the city. The city is implemented stage by stage, and once a phase is completed, the area of Masdar City that corresponds to that phase becomes immediately operative. From an economic perspective, this means that the business partners of the Masdar Initiative do not have to wait decades to move to Masdar City. They can establish their offices and laboratories in the operating zones of Masdar City while the rest of the Emirati city is under construction. As a result, the Masdarian business based on the development and commercialization of clean technologies can be quickly put into practice, and revenues can be generated from the first stage of the project.

In a similar vein, the development of on-site energy sources takes place following modular patterns. Every area of Masdar City is designed to be supported by a series of technologies generating clean energy. Roof-mounted photovoltaics, concentrated solar power and geothermal stations are examples of the Masdarian technology. The aim of the Masdar Initiative is to develop energy-independent buildings that can be inhabited when large portions of the city are still under construction. Ultimately, Masdar City is meant to be a large-scale power source whose energy gradually increases stage by stage. By adopting a modular type of planning, the Masdar Initiative is able to activate the energy potential of Masdar City not at the end, but at the beginning of the project (although see Cugurullo 2015).

'Fast' regulation

The speed that characterizes the development of the Masdar City project can be further observed and explained by looking at the building regulatory system that frames the implementation of the new city. Estidama is the name of the local system designed to regulate and assess the built environment of Abu Dhabi. It is described by the Abu Dhabi Urban Planning Council (UPC) as a 'framework for measuring sustainability' from 'four predefined angles: environmental, economic, social and cultural' (UPC 2014). The core element of Estidama is the Pearl Rating System (PRS) that provides a series of parameters, such as carbon emissions and energy waste, through which to assess the sustainability of Abu Dhabi's urban environments. The mechanism is simple. Every time an urban project meets the requirements of Estidama, the UPC gives it a 'Pearl'. The PRS is compulsory and, to be approved, a project has to reach a specific minimum that varies according to the scale and political nature of the project: one Pearl over five for new buildings and communities, and two Pearls for government-led projects. What is peculiar about Estidama is its origin. Interview data showed that Estidama and Masdar City started to be developed at the same time and by the same group of actors. It was the Masdar Initiative itself that designed the mechanics of Estidama while working on the mechanics of the Masdar City project. As a result, it is no surprise that what Estidama measures mirrors the characteristics that Masdar City possesses. The Masdar City project was quickly approved by the UPC and, in early 2010, the entire city received two Pearls. Back then, only 15 per cent of the project had been developed and Abu Dhabi's alleged eco-city consisted mostly of a large, dusty construction site. Nonetheless, for the local planning council Masdar City had fully met the requirements of Estidama, and the Masdarian case was closed.

The involvement of the UPC during the development of Masdar City is revelatory of how, since the genesis of the project, Abu Dhabi has accurately prevented any potential delay. To date, the role of the council has been limited to what a representative from the UPC described in an interview as 'conversations with the Masdar Initiative', and to sporadic examinations of master plans. The representative admitted that the entire project has been evaluated exclusively on the basis of planning documents provided by the developers. No *in situ* empirical examination has been made, and the council has no empirical data on the actual environmental performance of the new settlement. Nonetheless, the UPC has never asked the Masdar Initiative to produce any type of environmental data and, across the years, it has approved every draft of the master plan, giving a perpetual go-ahead to the Masdar City project.

Conclusions

Using Masdar City as a case study, this chapter has explored how, in the Global South, the concept of the eco-city has been integrated in the genesis and development of a new master-planned city. In the Masdar City project, the eco-city ideal is interpreted from an economic perspective and translated into a large-scale business project in which new urban spaces are designed and employed to generate economic returns. The aim of Masdar City is to be an urban laboratory and showroom where clean-tech products are developed and commercialized. Multinationals such as Siemens, Schneider and Mitsubishi are invited to join the eco-city project to increase the number and scope of the technologies produced in and by Masdar City, and their role is regulated via partnerships that have a substantial impact on the implementation of the master plan. As shown in the previous sections, the entire process of city-making underpinning Masdar City is shaped by the economic interests of the Masdar Initiative and its business partners. The master plan of the new city does not follow an overarching planning and architectural vision, and the original ideas of space envisioned by Foster and Partners have been changed over the years to support the formation of new business partnerships.

In this chapter, the concepts of speed and fast urbanism have put emphasis on the velocity that characterizes the Masdar City project. The chapter has shown that in the process of city-building, speed comes at a price. There is a price to pay when urban developers decide to accelerate the implementation of a new city and, in the case of an eco-city project like Masdar City, the price has been *sustainability*. Masdar City is a 'fast' city. Its development is one of the top priorities of Abu Dhabi whose political elites want the project to be fully implemented and operative before the end of the Oil Age. Consequently, the Masdar Initiative has to opt for rapid planning strategies that, by favouring the formation of business partnerships, penalize the cultivation of a homogenous master plan driven by a precise vision of the city. In addition, Masdar City is framed by an *ad hoc* building regulation that has been tailored specifically around the Emirati eco-city project. Via the local planning council and the Masdar Initiative, Abu Dhabi has created a building regulatory system, Estidama, which further increments the speed of implementation.

However, city-making is a complex process and being fast means neglecting some of its components. With regards to sustainability, the Masdarian attitude towards city-making is highly problematic. First and foremost, the focus of the Masdar Initiative on the economic aspects of the city is automatically translated into a disregard for its environmental and social dimensions. The environmental dimension of Masdar City is cultivated to boost the clean-tech business of the Masdar Initiative: a business that, by focusing on technologies designed to achieve carbon neutrality, pays little or no attention to other important environmental issues such as ecosystems and ecosystem services. In terms of the social dimension of the Emirati eco-city project, as noted in previous studies (see Cugurullo, 2013a, 2013b), the interests of stakeholders are minimum and the risk is the formation of a *non-place* in which social relationships, arguably the kernel of urban living, are replaced by business relationships. From this point of view, Masdar City appears to be understood by its developers as a business project rather than as an urban and therefore social settlement. In this sense, the concerns raised by this chapter are not about what type of society Masdar City will shape, but if a society will be developed at all.

Understanding Masdar City means understanding part of the urban future of the Global South. Since the early stages of the project, the Masdar Initiative has been keen to promote its eco-city model as a universal paradigm of city-making. The new city has often been described by developers and stakeholders as a model of the city of the future in the attempt to establish the Masdarian urban vision outside Abu Dhabi. Over the last seven years, the Emirati project has gained considerable visibility, featuring in textbooks of planning as well as in international reports as an example of urban sustainability (see, for instance, Lehman 2010, UN Habitat 2009). Moreover, interview data suggests the presence of a broad and solid policy network linking Masdar City with a number of urban projects from the Global South. According to a representative from the Masdar Initiative, Abu Dhabi has already established several collaborations with local urban initiatives in South-East Asia: collaborations that see members of the Masdar Initiative working *in situ* to mentor developers and policy makers eager to replicate the Masdarian vision in their country.

Although it is unlikely that the Emirati eco-city model will be perfectly replicated in different geographical contexts, the Masdarian vision of the city has the power to impact on the way new cities are envisioned and implemented. This chapter has attempted to open a window into the urban future that Masdar City is shaping. The Masdarian example raises concerns over how the concept of the eco-city has been integrated in the urbanization of the Global South. The model of city-making promoted by the Masdar Initiative sacrifices important environmental and social aspects of urbanism to quickly satisfy broader economic and political ambitions. This is a potentially dangerous road that can lead developing countries to experience major cases of environmental degradation and social injustice in the near future. Today, the Global South is rapidly urbanizing and *today* is when such urbanization has to be critically examined and evaluated, before its environmental and social impact becomes irreversible.

Note

1 New eco-city projects of the Global South: Masdar City (United Arab Emirates), Nanjing Eco High-Tech Island, Yinggehai low-carbon city, Tianjin Eco-city, Xinjin Water City, Caofeidian, Wanzhuang (China), Changodar, Dahej, Manesar, Bawal, Shendra (India), Sseesamirembe Eco-city (Uganda), Zenata Eco-city (Morocco), Hanoi Green City (Vietnam), Ecobay (Estonia), Zira Island (Azerbaijan), Rawabi (Palestine), Punggol Eco-Town (Singapore), Blue City (Oman), K.A.CARE (Saudi Arabia).

References

Abu Dhabi Government. 2009. *Economic Vision 2030*. 3 January. Accessed on 3 January 2008 from https://www.ecouncil.ae/PublicationsEn/plan-abu-dhabi-full-version-EN.pdf.

Abu Dhabi Urban Planning Council. 2014. *Estidama*. Accessed on 3 January 2014 from http://estidama.upc.gov.ae/pearl-rating-system-v10/pearl-community-rating system.aspx.

Brand, P. and Thomas, M.J. 2005. *Urban Environmentalism: Global Change and the Mediation of Conflict*. London: Routledge.

Chang, I.C. and Shepherd, E. 2013. "China's eco-cities as variegated urban sustainability: Dongtan Eco-City and Chongming Eco-Island." *Journal of Urban Technology* 20: 57–75.

Cugurullo, F. 2013a. "How to build a sandcastle: an analysis of the genesis and development of Masdar City." *Journal of Urban Technology* 20: 23–37.

Cugurullo, F. 2013b. "The business of utopia: Estidma and the road to the sustainable city." *Utopian Studies* 24: 66–88.

Cugurullo, F. 2015. "Urban eco-modernization and the policy context of new eco-city projects: where Masdar City fails and why." *Urban Studies* Advance online publication. doi: 10.1177/0042098015588727.

Datta, A. 2012. "India's ecocity? Environment, urbanization and mobility in the making of Lavassa." *Environment and Planning* 30: 982–996.

Elkington, J. 2004. "Enter the triple bottom line." In *The Triple Bottom Line: Does it All Add Up?*, by A. Henriques and J. Richardson (eds.), 1–16. London: Earthscan.

Freedom House. 2014. *Freedom in the World.* Accessed on 3 January 2014 from https://freedomhouse.org/report-types/freedom-world#.VeHBnPmqqko.

Joss, S. and Molella, A.P. 2013. "The eco-city as urban technology: perspectives on Caofedian International Eco-City (China)." *Journal of Urban Technology* 20: 115–137.

Kargon, R.H. and Mollela, A.P. 2008. *Invented Edens: Techno-Cities of the Twentieth Century.* Cambridge: The MIT Press.

Krueger, R. and Gibbs, D. (eds.). 2007. *The Sustainable Development Paradox: Urban Political Economy in the United States and Europe.* London: The Guilford Press.

Lehman, S. 2010. *The Principles of Green Urbanism: Transforming the City for Sustainability.* London: Earthscan.

Masdar Initiative. 2014. *About Masdar Capital.* Accessed on 3 January 2014 from. http://www.masdar.ae/en/investment/about-masdar-capital.

Merrifield, A. 2012. "The urban question under planetary urbanization." *International Journal of Urban and Regional Planning* 37: 909–922.

Rapoport, E. 2014. "Utopian visions and real estate dreams: the eco-city past, present and future." *Geography Encompass* 8: 137–149.

Shwayri, S.T. 2013. "A model Korean ubiquitous eco-city? The politics of making Songdo." *Journal of Urban Technology* 20: 39–55.

UN Habitat. 2009. *Planning Sustainable Cities: Global Report on Human Settlements.* London: Earthscan.

Urban Age Project. 2007. *The Endless City: The Urban Age Project.* Phaidon, London: The London School of Economics and Deutsche Bank's Alfred Herrhausen Society.

——. 2011. *Living in the Endless City: The Urban Age Project.* Phaidon, London: The London School of Economics and Deutsche Bank's Alfred Herrhaussen Society.

WCED. 1987. *Our Common Future.* Oxford: Oxford University Press.

Part II

ENTREPRENEURIAL STATES

5

ENVISIONED BY THE STATE

Entrepreneurial urbanism and the making of Songdo City, South Korea

Hyun Bang Shin

Introduction

So much has been said about Songdo City in recent years in both academic and practitioner circles. International media has also taken part in inflating the reputation of Songdo City, hailed initially as an eco-city, then as a ubiquitous city (or U-city) and now a smart city (Kim 2010, Shin et al. forthcoming, Shwayri 2013). The *New York Times* went even further to dub it "Korea's High-Tech Utopia" (O'Connell 2005). Sometimes its own promotional material puts all these together and simply refers to Songdo as an eco-friendly ubiquitous smart city (IFEZ Authority 2007). Governments elsewhere see Songdo as a reference for their own mega-projects to create a brand new city from scratch (see *El Telégrafo* 2012 for an example of the construction of Yachay City in Ecuador). However, Songdo has come to cater exclusively for the needs of domestic and global investors as well as the rich who have financial resources to grab upmarket real estate properties. It may indeed be an urban utopia, built on a reclaimed *tabula rasa* and promoted by the state, merging together technological innovation, fixed assets investment, real estate speculation and financialization, for the exclusive use of the rich and powerful.

In her examination of smart city frenzy in India, Ayona Datta (2015) highlights the extent to which smart city promotion is the extension of "a longer genealogy of utopian urban planning" in the country, and how smart city construction has been part of entrepreneurial urbanization that positions new city promotion as a lynchpin for economic growth. The review of the development history of Songdo City in this chapter echoes the experience of India to some extent. While Songdo City is often regarded as an emblem of a new global rhetoric of technological fixes for urban problems, its promotion is deeply rooted in the history of territorial planning by the South Korean developmental state that has been behind the rise of spatial fixes, speculative urbanization and vertical accumulation (Harvey 1978, Shin 2011, Shin and Kim 2016). Furthermore, this chapter ascertains that despite all the rhetoric surrounding Songdo City, it represents the territorial manifestation of the legacy of the Korean developmental state that has been internalizing the neoliberal logics of capital accumulation and by doing so, sustaining its presence in city-making (Choi 2011, Part et al. 2012). The construction of Songdo City has benefited from the country's liberalizing financial system for project financing and also from the public–private partnership between the local state and domestic and transnational firms. While the presence of transnational investors is often highlighted by the media, endogenous institutional landscape and spatial practices turn out to be more sticky and resilient vis-à-vis pressures of neoliberal urbanization (Peck et al. 2009). The history of Songdo City also demonstrates that the growth promoters that comprise local and central states as well as domestic businesses partake in scalar politics (see also Shin et al. forthcoming), creating a *tabula rasa* for the realization of their ambition to maximize gains from city marketing and accumulation of real estate capital.

The rest of this chapter examines (i) the emergence of local state entrepreneurialism and its impact on city-making in South Korea (hereafter Korea), (ii) the history of making Songdo City and (iii) the role of key institutions

Figure 5.1 Entering Songdo City (photographed by Hyun Bang Shin, 2011)

such as domestic financial institutions. These discussions are followed by a scrutiny of how the construction of Songdo City depends heavily on extracting value from real estate, which has been a key characteristic of urban development under the Korean developmental state for many decades. The arguments in this chapter make use of primary and secondary data collected from a variety of sources including the author's interviews with key informants conducted in December 2011, media reports, corporate reports and government archives. To understand the historical evolution of the project and the associated contentious developmental politics, this chapter adopts multi-scalar approaches to locate the different roles played by transnational, national and local players who have contributed to the rise of Songdo City in their own right.

Legacy of developmental state and entrepreneurialism

East Asian states have been known for their developmentalism, which guided the region's condensed urbanization and industrialization in the second half of the 20th century. Developmental states are characterized by their proactive intervention in governing the market, nurturing human capital, disciplining the workforce and subordinating social policies to economic policies (Woo-Cumings 1999). They have also concentrated finite resources on economic development, mobilizing national savings as well as foreign loans and/or direct investments to make productive investments in the built environment (Harvey 1978, Shin and Kim 2016). In the case of Korea, the state was forming an alliance with large conglomerates (often known as *Chaebol* in Korean) to nurture strategically prioritized industries, subsidizing their businesses with preferential access to foreign loans and making use of state companies to provide infrastructure and facilities (Shin 2007).

With the neoliberalization pressure and the political decentralization of the Korean state from the 1990s (see Pirie 2008), the central and local states have been resorting more to private-led initiatives. The Asian financial crisis in the late 1990s also forced the central state in particular to reconfigure its regulatory arrangements and the ways in which it intervened in regional development. Furthermore, from the mid-1990s, the rise of local state entrepreneurialism has become evident, thanks to the political decentralization that resulted in the direct election of mayors and provincial governors (Choi 2011). Entrepreneurial cities are characterized by a dependence on domestic and global capital for local economic development, and the promotion of urban development led by mega-projects and spectacular landmark buildings to create investor-friendly landscapes: in short, creative destruction to rebuild cities to channel surplus capital into the built environment and produce speculative profits (see also Hall and Hubbard 1998, Harvey 1989, Shin 2009). The presence of a strong authoritarian developmental state before the 1990s did not rule out the rise of local growth politics (see, for example,

Park 2008), but it has become fiercer after political decentralization, as directly elected mayors and provincial governors propagate boosterish policies to attract investment as well as central government subsidies.

Incheon, which administers Songdo City, has been mostly dominated by those mayors who personified their political ambition in transforming Incheon into a city of construction and speculation. Most mayors of Incheon until 2010 were loyal to the right-wing political faction that held the power and resources to sustain Korea's authoritarian developmental statism between the 1960s and 1990s. For instance, Mayor Gi-seon Choi had been an appointed mayor between March 1993 and September 1994, nominated as a mayoral candidate by the then ruling right-wing party in 1995, and became the first elected mayor, serving two terms until 2002 after successfully running for re-election. His election pledge in the 1995 election was to promote development projects to raise surplus that could be reinvested in Incheon's infrastructure provision (The Dong-A Ilbo 1995), and he remained committed to this pledge. Mayor Sang-soo Ahn, who was the elected mayor between 2002 and 2010, was from the same party as his predecessor, carrying on and even expanding the boosterish policies during his terms of office.

Incheon's entrepreneurial policies resulted in its successful bid for the 2014 Summer Asian Games in April 2007, and led to a rapid expansion of state-led urban redevelopment projects that have grown in scale in recent years. However, these came at the expense of fiscal stability for Incheon (Shin 2016a). Already in 2000, concerns were raised about the excessive debts the city was incurring to bear the costs of land reclamation. As of 2000, municipal bonds reaching the value of 150 billion Korean Won[1] were issued to pay for land reclamation, and the possibility of the city going bankrupt was looming when the city began repaying the principal and interests from 2004 (The Dong-A Ilbo 2000). Incheon's fiscal independence was relatively high (62.6 per cent in 2014) compared to other municipalities (with the exception of Seoul, which enjoyed the highest fiscal independence among all cities in Korea) but substantially undermined when compared to the level in 2004 (75.9 per cent).[2] The total debts of the municipality amounted to 37.7 per cent of its annual budget in 2011, considerably higher than the average of all cities and provinces (21.0 per cent).[3] The deteriorating fiscal conditions in turn seem to have reinforced the speculative and entrepreneurial nature of Songdo City development, creating more opportunities for increasing the share of residential flats, which provided a greater chance of raising larger revenues than commercial or business premises.

The rise of the local state entrepreneurialism should be examined in a more nuanced way, as the central state remains powerful. The day-to-day running of affairs within the Incheon Free Economic Zone (hereafter Incheon FEZ) is governed by the Incheon FEZ Authority, which is in turn controlled by the Incheon municipality. However, the central government's

FEZ Committee wields greater power over all the FEZs in Korea, "regulating it both spatially and socially through drawing up and implementing development plans, approving master plans, and designing tax benefits for individuals and businesses" (Shwayri 2013, 46). Furthermore, local governments' poor fiscal capacity means that their future is predicated upon the size of central government subsidies.

Birth of a new city and the Korean state

Songdo City is located about 50 kilometers from the center of Seoul, the Korean capital of culture, economy and politics. Songdo City is also part of the Incheon FEZ. The Incheon FEZ was born out of a massive reclamation project to create a new land mass of 169.5 km². The first phase of Incheon FEZ development involving reclamation and infrastructure provision was completed in 2009. The final completion of the entire Incheon FEZ construction is expected by 2020. Songdo City occupies about one third (31.5 per cent or 53.4 km²) of the Incheon FEZ. Songdo City proclaims to be an "aerotropolis," benefiting from its proximity to the country's main international gate, the Incheon International Airport built on the neighboring Yeongjong City of the Incheon FEZ. The proximity to Seoul and the international airport has been a key geographical advantage that the Incheon FEZ and developers of Songdo City frequently publicize.

Contrary to the fairly recent international attention to the rise of Songdo City, the idea of building a brand new city from scratch and creating a city of nature, information technology and an international center of trade is deeply rooted in the country's territorial planning history, which spans across several decades. First, the idea of building Songdo City on reclaimed land was part of Korea's national development planning. Like in Hong Kong or Singapore, land reclamation to expand the national territory has long characterized Korea's territorial development strategy. The idea of reclaiming the sea around Songdo Island (hence the origin of the name Songdo City) could be traced as far back as 1962 when a private company, Woojin Mulsan, had an ambitious plan to reclaim about 5 km² to build a brand new city, which was approved by the central government on 30 December 1961. The land reclamation was to be privately financed by the company, and according to the then Land Reclamation Act, the company would own the reclaimed land (The KyungHyang Shinmun 1962).

The above plan seems to have been overly ambitious, as its progress cannot be traced in any of the national media archives, but the idea of building a brand new city on reclaimed land re-emerged in April 1988. Upon the visit of the then President Roh Tae-Woo to Incheon, the municipality gave a briefing about their plan to reclaim 48 km² around Songdo Island to create an international center of trade and information technology (The Hankyoreh 1993). Having further concretized the plan, the municipality proclaimed in April 1991 its plan to reclaim 17.7 km² during the next five years. The reclaimed land was to accommodate the construction of 70,000 housing units and a central business district of information technology, communication and commerce (Maeil Business 1991). Incheon, having experienced a shortage of land supply, hoped to secure a new supply of land to build houses and also use revenues through land sales and lease to finance infrastructure provision in existing urban built-up areas (Maeil Business 1989). Approvals were gained from the Ministry of Construction in the central government to proceed with the reclamation and design of urban master plans in 1990 and 1991 respectively. After a short delay of three years due to the central government-led construction of a new international airport nearby, the reclamation of Songdo Island eventually started on 10 September 1994 (The Hankyoreh 1993, 1994).[4]

The municipality was eager to produce a "city of the future". In December 1997, it signed a memorandum of understanding with a consortium of 17 private firms to establish a dedicated district (3.5 km²) to establish a "knowledge information industrial complex" to accommodate IT firms and create what would hopefully become a Korean version of "Silicon Valley" (The Dong-A Ilbo 1997, Maeil Business 1997a). While doubts were raised about how the municipality was going to finance the massive construction costs, the municipality remained optimistic that private firms would be willing to pay for the purchase or lease of lands when the lands were ready for sale by October 1998 (The Dong-A Ilbo 1997). It was also announced that an international business district (6.8 km² – larger than what was eventually finalized) would be created for finance and business, including convention facilities as well as residential buildings (Maeil Business 1997a).

All these plans turned out to be too optimistic and were substantially delayed when the country was severely hit by the Asian financial crisis from late 1997. The Incheon municipality was not able to raise enough revenues from the sale of reclaimed lands due to the downturn of the real estate market in the immediate aftermath of the crisis. It was also reported in mid-February 1998 that the resource-stricken municipality only secured 38 per cent of its annual budget for the reclamation to be carried out in 1998 (The Hankyoreh 1998).

The resumption of Songdo City development during the post-crisis recovery period was particularly helped by the central government policy to create FEZs and attract foreign direct investment (FDI). The establishment of these FEZs was part of the manifesto of the President Kim Dae-jung administration (1998–2003) that aimed to steer the Korean economy away from being too dependent on manufacturing industry and towards an economy of international logistics, finance and high-tech industries including information technology and electronics. Subsequently, the national government put forward its economic development vision to place Korea as a northeast Asian business hub. In April 2002, a blueprint was further announced, part of which included the rising importance of major port cities for international logistics: land reclamation was of particular importance to carry out comprehensive development and accommodate those industries and facilities that the governments desired. On 30 December 2002, the Act on Designation and Operation of Free Economic Zones was enacted, and was enforced from 1 July 2003. Initially, three port cities, Incheon, Busan and Gwangyang, came to have FEZs within their jurisdiction. In Incheon, three land reclamation areas came to constitute the Incheon FEZ: Songdo City, Yeongjong City (where the new international airport is located) and Cheongna City.

It was hoped these FEZs would "lur[e] multinational companies" so that the country would "win back the foreign direct investment lost to China and others over the past four years" (Song 2004). To make this possible, FEZs were to become places "where no tariffs are imposed on capital goods and only 17 per cent income tax is levied on foreign corporate executives and cash grants are offered to foreign companies building high-tech facilities" (Song 2004). Retrospectively, such desire seemed to have been very ambitious, especially given the exponential growth of China's receipt of FDI throughout the 2000s and given Incheon's proximity to Shanghai. Nevertheless, the strategic importance of Incheon FEZ grew when the President Roh Moo-hyun administration (2003–8) inherited the previous administration's emphasis on positioning Korea as the "central state" of north-east Asian economies. The production of the central state's national development strategy to promote the Incheon FEZ provided the necessary momentum Songdo City desperately needed, including the central state's financial undertaking for some of the key infrastructure projects.

Pivotal role of domestic players in building a city of the future

Songdo City's international popularity was boosted by the construction of Songdo International Business District (hereafter Songdo IBD) on a piece of reclaimed land that was sold to a joint venture whose majority share was held by a US real estate developer, Gale International. The joint venture possessed exclusive development rights for the construction of the Songdo IBD, bringing Songdo City close to what Gavin Shatkin (2011) coined as "privatopolis". It was the joint venture that heavily promoted Songdo City's eco and smart city brand from the outset. For instance, in August 2001, Gale International appointed Kohn Petersen Fox (KPF) as the architect for master planning of the Songdo IBD in August 2001, and the KPF chose a sustainable city (or an eco-city) approach for the design (Segel 2012, 7).[5] The then CEO of Gale International reflected on the early period in an interview that his company had to make a considerable effort to persuade the Korean government to place greater value on "quality-of-life projects like the park and not for residential or office buildings," and prided himself that this approach was "a clear example where we as 'foreign' developers really brought an international perspective to this development" (cited in Segel 2012, 6). Songdo IBD now sells itself on its website as being "one of the world's greenest cities," built on six design goals that aim to reduce carbon emission and waste, promote ample open and green spaces, and increase energy efficiency. For instance, the transnational stakeholders, Gale International and CISCO in particular, celebrated their joint effort to replicate the Songdo IBD experience in China's Meixi Lake development in Changsha, Hunan Province.[6] However, while Songdo City aspires to become transnational in its outreach, the process of developing Songdo City has been very much "Korean". Two aspects are discussed in what follows: (i) the pivotal role of domestic financial institutions and (ii) "green urbanism" as the Korean state's aspiration rather than imposition by transnational investors.

First, in the aftermath of the Asian financial crisis in the late 1990s, the Incheon municipality was hoping to attract as much foreign capital as possible, but the overall construction of Songdo City eventually came to depend heavily on domestic capital. The close scrutiny of the development process of Songdo IBD testifies to this. The construction of Songdo City was given additional momentum when a US–Korea joint venture signed an agreement with the Incheon municipality to develop the Songdo IBD, the most popularized 5.77-km² district within Songdo City. Initially, POSCO E&C, a subsidiary of the Korean steel giant POSCO, was granted a fixed term (six months) status as the lead master plan developer on the condition that its status would become permanent if it teamed up with an overseas developer (Kim and Ahn 2011, 663, Shwayri 2013, 46). Eventually, Gale International based in New York City was brought in, and upon signing a memorandum of understanding with the Incheon municipality in July 2001, the Gale-POSCO E&C partnership was granted exclusive rights to develop the business district.[7] Gale International and POSCO E&C set up a joint venture named New Songdo City Development (renamed in 2007 as New Songdo International City Development (NSIC)) dedicated to the Songdo IBD project with the 70 per cent majority share by Gale International.[8] The joint venture's award of the exclusive development rights and the government subsidies in the form of discounted land sales were conditional on the joint venture's attraction of FDI promised to be worth US$12.7 billion. The total project cost was expected to reach KRW 24.4 trillion (about US$18.4 billion), and almost all the costs (98.4 per cent) were to be borne by the NSIC (IFEZ Authority 2007, 176).

Instead of attracting FDI, however, the joint venture benefited heavily from the loans arranged by domestic financial institutions. The role of transnational capital, epitomized by the participation of Gale International and other investors such as CISCO and 3M, has been marginal in terms of making meaningful financial contributions. As of the end of 2008, the actual

FDI in the Songdo IBD remained minimal, reaching a mere 1.6 per cent (US$33.5 million) of the original plan (Chosun Ilbo 2010). The majority of development finance was mobilized by a consortium of Korean banks and other financial firms who sought a faction of profits from the real estate projects in the Songdo IBD. For instance, when the NSIC sought the initial finance of US$90 million to purchase the first lot (about 0.33 km²) of reclaimed land in October 2003, US$50 million was reported to have been secured from a consortium of Korean banks (Woori Bank and Industrial Bank of Korea), while the remaining US$40 million came from international investors including Morgan Stanley and ABN Amro (Munhwa Ilbo 2003, Segel 2012, 10). The role of Korean financial institutions continued to be influential when the NSIC sought additional loans to finance its business in subsequent years.[9]

Second, Songdo City prides itself on being an environmental and smart city, a key selling point in the global real estate market, but green urbanism was conceived by the Korean state as early as the late 1980s. Back then, the envisioned new city was to become a vertical city with high-density construction to release about 60 per cent of the land for recreation facilities and greenery so that it could replicate the strengths of cities like Singapore and Canberra (The Hankyoreh 1993). When a plan was announced in April 1991 for the ambitious land reclamation, it also aimed to create a new town by 1996, which would have 35 per cent of its surface area dedicated to parks and greenery (Maeil Business 1991). Korea also witnessed the proliferation of national development strategies such as cyber-Korea and e-Korea that tried to integrate information and communication technologies in transforming the ways in which people lived and worked (Yigitcanlar and Lee 2014). By the time the master plan of Songdo IBD was finalized, "[g]overnment officials envisage[d] an 'eco-friendly and intelligent' city completely different from Seoul, which has become an overcrowded concrete jungle after decades of rapid industrial development. Songdo is planned as a waterfront city laced with 6km of canals, a central park, a golf course and pollution-free transport" (Song 2004). Such views of government officials might have been the outcome of the strong persuasion of Gale International, but the idea of building a new city with ample green space and equipped with information technology was clearly not unknown to the Korean state. Songdo City's branding as an eco-city was further boosted by the central state's emphasis on "green growth," a slogan that epitomized the national development policies by then President Lee Myung-bak (2008–13) (Kim 2010). Having faced the urgent need to overcome the detrimental impact of the global financial crisis on Korea, the Lee administration produced a series of controversial economic stimuli programmes that particularly emphasized large-scale construction activities to rewrite the territorial landscape under the rubric of "green investments". Songdo City fit nicely into this "green" endeavour.

Extracting value from real estate

While Songdo City gained its fame as an environmental and smart city, its construction depended heavily on real estate investment. In short, Songdo City was the site of real estate speculation, involving both domestic and transnational capital. The symbolic aestheticization of Songdo City (Kim 2010) appears to have aimed ultimately at marketing the real estate project. This was particularly pronounced during the post-Asian financial crisis period that saw heightened speculative real estate markets after an initial slump (see also Goldman 2011, Shin and Kim 2016 on the rise of speculative urbanization). Lenders, most of whom comprised Korean firms (banks, insurance and security firms), betted on the residential and retail components of the development. For instance, when the NSIC was taking out loans in 2005, it was noted that the Korean banks (and ABN Amro) "were underwriting this loan based solely on revenues from the residential and retail buildings" despite the Korean government's emphasis on office development (Segel 2012, 11).

Songdo City developers benefited from the huge subsidies provided by the municipal government in the form of cheap prices on reclaimed lands, a practice that is somewhat reminiscent of mainland China (Hsing 2010, Shin 2016b). Developers would have been able to take advantage of rapidly increasing land prices as collateral when negotiating with financial institutions for loans. For instance, government records suggest that the official land price (evaluated for tax purposes) for the site of Songdo City's landmark buildings (the Northeast Asia Trade Tower and Songdo Convensia) was only KRW 160,000/m^2 in 2002, the year when the Incheon municipality and NSIC were concluding the land supply contract (IFEZ Authority 2007, 176).[10] The land price rose sharply to KRW 1,110,000/m^2 in 2003, KRW 1,500,000 /m^2 in 2004, KRW 1,800,000/m^2 in 2005, and KRW 4,200,000/m^2 in 2007. Here, the land price data are presented for those years that the NSIC took out loans. According to the report from the Citizens' Coalition for Economic Justice (2010, 5), an influential civil society organization in Korea, the NSIC completed the purchase of 3.3 km^2 by the end of July 2009, having paid KRW 866.7 billion. This would result in an average sale price of KRW 260,000/m^2, making it clear that the joint venture would have seen windfall profits by taking advantage of the difference between the actual purchase price of land and the substantially increased official land price.

Furthermore, the sale of residential units was central to project financing. In the case of Songdo IBD, only half of the originally estimated building volumes were completed as of December 2013, with the anticipated completion year postponed to as late as 2018 (Nam 2013). Under the circumstances, selling real estate properties and finding tenants were key to the financial viability of the project. As the demand for offices and retail spaces was lower than expected and that for residential space much higher (and hence more likely to produce higher profits), the land use plan was revised so that the size of land allocated to international business use was reduced by 38.4 per cent between February 2004 and August 2009, while that of residential complexes increased by 24 per cent during the same period (Citizens' Coalition for Economic Justice 2010). The NSIC sold in total 5,364 condominium units between 2005 and 2009: their average size was 152 m² and the average sale price, KRW 4,132,150/m². Each unit was therefore sold at the hefty average price of KRW 628 million, which would amount to about 15 times the average annual household income for urban households in 2009.[11] The extortionate price suggested that only the most affluent in society with purchasing power were to consume the "quality of life" offered by the Songdo IBD, and that the NSIC developers would have raised a large amount of profits from the sale of these residential units. It was estimated that about KRW 1.9 trillion (about US$1.6 billion)[12] would have been raised by the joint venture as profits from the sale of these residential units, after taking into account all the costs of land purchase, site preparation, construction, finance and taxation (Citizens' Coalition for Economic Justice 2010). The significance of constructing residential complexes in Songdo City is unlikely to diminish, given the report that "about half of the commercial space in Songdo's gleaming office skyscrapers is empty because the business district has struggled to attract tenants" as of December 2013 (Nam 2013). Such situations make the Songdo City development even more speculative, as the fate of the city depends entirely on the market conditions.

Conclusion

To some extent, the development of Songdo City fits quite well with the experience of post-industrial cities in the global North that have tried to survive economic misfortune in recent decades. Incheon has lost much of its industrial basis to other Asian countries, after having lost the previous advantages of low-cost labor disciplined by the authoritarian state (see Arrighi 2009 for the snowballing of labor-intensive industries within East and Southeast Asia). The public–private partnership in the construction of Songdo City testifies to the strong element of urban entrepreneurialism as identified by David Harvey (1989) in his discussion of governance transition in post-industrial cities like Baltimore in the US. First, there is a public–private partnership that uses the state power to entice external sources of funding, especially global capital (Harvey 1989, 7). The project is also highly "speculative in execution and design" (ibid.), especially when reaching the most recent stage of promoting the Songdo IBD. The development of Songdo City rests on "the political economy of place rather than of territory" (ibid.). That is, the project is centered on the generation of exchange value through speculative promotion of its real estate components rather than paying equal attention to the provision of collective consumption for all strata of the population.

Furthermore, Songdo City, being part of the Incheon FEZ, has become a "zone of exception" (Agamben 2005, Ong 2006, Park 2005, Wu and Phelps 2011) where rules are bent and preferential treatments are offered to a group of privileged stakeholders. Local and central states as well as businesses work together to transcend the geographically bound developmental politics by means of creating new zones of exception through land reclamation and/or designation of FEZs, overcoming the constraints in existing urban spaces (e.g. resistance by local residents against displacement) to fast-track the developmental aspiration of the state and capital. The site has become part of producing "elite urban spaces" (Shatkin 2011, 83), an off-limit space from any public disorder that disrupts new forms of city-making. The site may also be part of experimenting with various neoliberal policies of privatizing public services, which have not been applied elsewhere in the country previously, raising fears among the more progressive sectors of civil society with regard to the future expansion of such schemes at a national scale.

However, this chapter also shows that the story of Songdo City is not simply a neoliberal imposition of a particular brand of global urbanism on a locality. Along with other state-led territorial projects, the construction of Songdo City strongly reflects the deeply rooted aspiration of the Korean developmental state "to attract capital and people (of the right sort)" by means of "[i]magining a city through the organization of spectacular urban spaces" (Harvey 1990, 92). As Shwayri (2013, 40) observes, "[t]he planning of Songdo, like the planned modern cities of the twentieth century, has ignored local realities while focusing on creating an exportable model, but what has emerged is a city and/or suburb that is exclusively Korean," thus resulting in Songdo City's "Koreanization" (ibid., 50). The former CEO of Gale International, the US partner of the aforementioned joint venture, is also reported as having admitted that Songdo IBD "basically became a Korean project with an American name only at the wheel" (Nam 2013). These comments imply the strong resilience of the Korean developmental urbanism.

Therefore, the construction of Songdo City (and in particular the Songdo IBD) demonstrates the ways in which the Korean developmental state has internalized the neoliberal logics of capital accumulation, market fundamentalism and inter-city competition for investment and skilled labor to sustain its presence in city-making. To this extent, Songdo City is indeed what Bae-Gyoon Park (2005, 855) refers to as "hybrid spaces composed of neo-liberalization and the existing regulatory strategies" of the Korean developmental state. The development of Songdo City is influenced largely by the developmental vision promoted by the local state (Incheon municipality), while the central state also advocates the project to address its own national development agenda to raise the country's geopolitical position. In Songdo City, green and smart urbanisms have conjoined to produce entrepreneurial and speculative urbanization (Datta 2015, Shin and Kim 2016) that centers on real estate speculation and state-led investment in the built environment, which are the key characteristics of developmentalist urbanization in Korea. Songdo City is likely to serve the interests of urban and national governments as well as domestic businesses more than it serves overseas private capital. In short, the experience of Songdo City promotion manifests a particular practice of city-making, which is deeply rooted in the actions of the Korean developmental state that promotes its own version of globalization.

What does Songdo City mean to the Korean people, and what kind of city will it eventually become? The stakeholders in Songdo City may strive to replicate the model at a larger geographical scale within and outside the national territory. Although the developers of Songdo IBD were preaching that "[b]y employing the best practices of urban planning and sustainable design, Songdo IBD offers residents, workers and visitors an unparalleled quality of life" (Songdo IBD website), such quality of life is clearly intended for consumption by urban elites and a small segment of the local population who have the right level of purchasing power. That is, those citizens and expats who may be deemed appropriate by the state, exercising "graduated sovereignty" as identified by Aihwa Ong (2000) and Bae-Gyoon Park (2005). For a new city constructed from scratch, there is also a dissonance with the rest of the country that lies outside the exclusive zone. In fact, Songdo City comes to display characteristics that are closer to what Douglass and Huang (2007) referred to as an "exclusive utopia for an emerging urban upper middle class" (1) where there is no democratic civic input (see also Caprotti 2014). This comes at the expense of no adequate attention to urban diversity and especially the "right to the urban" (Shin 2014) for working-class people who carry out the day-to-day maintenance and remain invisible, and for those increasingly marginalized, disempowered and disenfranchised without having the resources, networks and power to access what is being created as eco- and smart urban spaces (Marcuse 2011). In the foreseeable future, Songdo City will continue to become a segregated and exclusive space, catering for the needs of the rich and powerful and becoming their own version of an urban utopia. Nevertheless, the longevity of this exclusive urban landscape is subject to speculation, for the heavy reliance of Songdo City on real estate investment will be its own weakness when the Korean real estate market slumps.

Acknowledgements

The author acknowledges funding support from the National Research Foundation of Korea Grant funded by the Korean government (NRF-2014 S1A3A2044551). The author also thanks Soojin Chung and Do Young Oh for their assistance with interview arrangements and data collection respectively. Many thanks are also due to Ayona Datta and Abdul Shaban for their constructive editorial suggestions. The usual disclaimer applies.

Notes

1 As per the mid-market rate as of 31 December 2001, US$1 was equal to 1322.9 Korean Won. See "Current and Historical Rate Tables," XE, accessed 9 September 2015: http://www.xe.com/currencytables.
2 The fiscal independence data come from http://lofin.moi.go.kr/portal/main.do, a central government website for data on local finance.
3 The data on municipal debts are from the Ministry of the Interior website, http://www.moi.go.kr/.
4 Plans for the economic activities Songdo City would accommodate were nevertheless zigzagging. In January 1996, Mayor Choi of Incheon announced a master plan for Incheon, which included an ambitious plan to expand the Songdo area from 17.7 km² to 95.9 km² to accommodate high-tech industrial complexes and other factories located in existing urban cores, and reduce the residential space substantially (Maeil Business 1996). However, this ambitious expansion of land reclamation met with resistance from the Ministry of Construction and Transportation in the central government (Maeil Business 1997b).
5 While the master plan itself was approved by the Incheon municipality in November 2002 and its subsequent finalization was in 2003, it nevertheless took many more years before the master plan was realized. The KPF referred to a number of cities that were deemed successful and assembled their landmark characteristics to create a master plan for Songdo IBD (Segel 2012). Therefore, Songdo IBD has become a kitsch amalgam of copied landmarks: "Central Park, the large 100-acre green space, which was modeled after New York City's Central Park, has already been completed. Besides the expansive park, Songdo takes inspiration from many other famous attractions from around the world. Songdo will also include Italianate canals, Savannah-style parks, Parisian boulevards, and a convention center modeled after Jørn Utzon's iconic opera house" (Meinhold 2009). The KPF prides itself on having created the master plan, having been awarded the Urban Land Institute and Financial Times Sustainable Cities Award in 2008.
6 The international partners of the Songdo IBD did not obscure their intent to make use of the Songdo model as a prototype that could be exported. The Chief Globalization Office and Executive Vice President of CISCO Services openly asserted that, "Our collaboration on Songdo IBD with Gale International is a living example of the globally replicable model we are building for Smart+Connected Communities" (CISCO 2009). The then Chairman of Gale International also responded to this by emphasizing that, "We are globalizing the real estate industry by identifying and deploying integrated solutions in a unique replicable model" (ibid.).
7 See http://www.songdo.com/songdo-international-business-district/team/development-team.aspx (last accessed 9 September 2015).
8 See also http://history.poscoenc.com/housing/housing_04_02_02.htm (last accessed 9 September 2015).
9 In June 2004, bridge loans of US$180 million were provided by a consortium of banks involving Woori Bank as a major contributor (IFEZ Authority 2007, 176, Segel 2012, 10). In June 2005, the NSIC secured another US$1.5 billion in loans

to repay the previous bridge loan, finance land purchase and provide working capital, and the major providers again included Korean banks (Segel 2012, 11). In 2007, the joint venture secured US$2.7 billion in loans to buy an additional 1.2 km² of reclaimed land from the Incheon municipality and also to repay existing loans from the previous round of project financing: The additional loans were to be provided by a consortium of 13 Korean banks and life insurance companies, led by Shinhan Bank who alone provided US$1.7 billion (Ramstead 2007). In 2013, NSCI secured the fifth round of project financing whose value reached about US$2.2 billion, offered by a consortium of nine banks and securities firms led by the Korea Exchange Bank (Incheon Ilbo 2013).

10 See also the Korea Land Information System website, http://klis.incheon.go.kr/sis/main.do.

11 The average annual household income for urban households in 2009 was KRW 41,625,696. The income data come from the governmental website of the Korean Statistical Information Service, http://kosis.kr.

12 As per the mid-market rate as of 31 December 2009, US$1 was equal to 1170.5 Korean Won. See "Current and Historical Rate Tables," XE, accessed 9 September 2015: http://www.xe.com/currencytables.

References

Agamben, G. 2005. *State of Exception*. Chicago: The University of Chicago Press.

Arrighi, G. 2009. "China's market economy in the long-run." In Hung, H-F, ed., *China and the Transformation of Global Capitalism*. Baltimore, MD: Johns Hopkins University Press, 22–49.

Caprotti, F. 2014. "Critical research on eco-cities? A walk through the Sino-Singaporean Tianjin Eco-city." *Cities* 36: 10–17.

Choi, B-D. 2011. "Neoliberal urbanization and projects of entrepreneurial city." *Journal of the Economic Geographical Society of Korea* 14(3): 263–285.

Chosun Ilbo 2010. "Songdo International City, lost (Gil ileun 'Songdo gugje dosi')." 14 March. Accessed on 9 September 2015 from http://businessnews.chosun.com/site/data/html_dir/2010/03/15/2010031500192.html.

CISCO. 2009. "CISCO and Gale International build on success of smart-connected communities in Songdo International Business District." 12 August. Accessed on 9 September 2015 from http://newsroom.cisco.com/press-release-content?type=webcontent&articleId=5072127.

Citizens' Coalition for Economic Justice. 2010. "Estimation of development profits in Songdo International Business District (Songdo Gugje Eopmu Danji Gaebal Yiyig Chujeong Balpyo)." 13 May. Accessed on 9 September 2015 from http://www.ccej.or.kr/index.php?document_srl=130376 [in Korean].

Datta, A. 2015. "New urban utopias of postcolonial India: 'entrepreneurial urbanization' in Dholera smart city, Gujarat." *Dialogues in Human Geography* 5(1): 3–22.

Douglass, M. and Huang, L.L. 2007. "Globalizing the city in Southeast Asia: utopia on the urban edge – the case of Phu My Hung, Saigon." *International Journal of Asia Pacific Studies* 3(2): 1–42.

El Telégrafo. 2012. "Corea del Sur brinda asesoría para el proyecto Yachay." 26 May. Accessed on 9 September 2015 from http://telegrafo.com.ec/sociedad/item/corea-del-sur-brinda-asesoria-para-el-proyecto-yachay.html [in Spanish].

Goldman, M. 2011. "Speculative urbanism and the making of the next world city." *International Journal of Urban and Regional Research* 35(3): 555–581.

Hall, T. and Hubbard, P. 1998. *The Entrepreneurial City: Geographies of Politics, Regime, and Representation*. New York: Wiley.

Harvey, D. 1978. "The urban process under capitalism: a framework for analysis." *International Journal of Urban and Regional Research* 2: 101–131.

——. 1989. "From managerialism to entrepreneurialism: the transformation in urban governance in late capitalism." *GeografiskaAnnaler. Series B, Human Geography* 71(1): 3–17.

——. 1990. *The Condition of Postmodernity: An Enquiry into the Origins of Cultural Change*. Oxford: Basil Blackwell.

Hsing, Y-T. 2010. *The Great Urban Transformation: Politics of Land and Property in China*. Oxford: Oxford University Press.

IFEZ (Incheon Free Economic Zone) Authority. 2007. "IFEZ White Paper: October 2005–Ocober 2007". Incheon: IFEZ Authority.

Incheon Ilbo. 2013. "Songdo IBD to receive a transfusion of 2.27 trillion won (Songdo IBD 2 jo 2700 eog suhyeol)." 17 December. Accessed on 9 September 2015 from http://m.incheonilbo.com/?mod=news&act=articleView&idxno=507567 [in Korean].

Kim, C. 2010. "Place promotion and symbolic characterization of New Songdo City, South Korea." *Cities* 27(1): 13–19.

Kim, J-W. and Ahn, Y-J. 2011. "Songdo Free Economic Zone in South Korea: a mega-project reflecting globalization?" *Journal of the Korean Geographical Society* 46(5): 662–672.

Maeil Business. 1989. "Living the 21st century in the ocean (21 segi en haesangseo sanda)." 11 October. 30 [in Korean].

——. 1991. "Commencement of the land reclamation from this October for the construction of Songdo new marine city (Songdo haesang sindosi geonseol ol siwol buteo maelip chagsu)." 9 April. 18 [in Korean].

——. 1996. "Cutting edge industrial complex on Songdo reclamation site (Songdo maelipji e cheomdan saneop giji)." 18 January. 37 [in Korean].

——. 1997a. "Land reclamation of 21.7 km² (655 man peyong gongyu sumyeon maelip)." 9 December. 37 [in Korean].

——. 1997b. "Painful construction of Songdo New City (Songdo sindosi joseong jintong)." 17 January. 37 [in Korean].

Marcuse, P. 2009. "From critical urban theory to the right to the city." *City: Analysis of Urban Trends, Culture, Theory, Policy, Action* 13(2–3): 185–197.

Meinhold, B. 2009. "Songdo IBD: South Korea's new eco-city." Inhabitat, 4 September. Accessed on 9 September 2015 from http://inhabitat.com/songdo-ibd-south-koreas-new-eco-city.

Munhwa Ilbo. 2003. "Exaggeration of the size of foreign capital attracted to Incheon special zone (Incheon teuggu oija yuchi 'bbeongtuigi')." 8 December [in Korean].

Nam, I-S. 2013. "South Korea's $35 billion 'labor of love': developer struggles to build a city from scratch." *The Wall Street Journal* (Online), 6 December.

O'Connell, P. 2005. "Korea's high-tech utopia, where everything is observed." *The New York Times*, 5 October. Accessed on 9 September 2015 from http://www.nytimes.com/2005/10/05/technology/techspecial/koreas-hightech-utopia-where-everything-is-observed.html?_r=0.

Ong, A. 2000. "Graduated sovereignty in south-east Asia." *Theory, Culture and Society* 17(4): 55–75.

——. 2006. *Neoliberalism as Exception: Mutations in Citizenship and Sovereignty*. Durham, NC: Duke University Press.

Park, B-G. 2005. "Spatially selective liberalization and graduated sovereignty: politics of neo-liberalism and 'special economic zones' in South Korea." *Political Geography* 24: 850–873.

——. 2008. "Uneven development, inter-scalar tensions, and the politics of decentralization in South Korea." *International Journal of Urban and Regional Research* 32(1): 40–59.

Park, B-G., Hill, R.C. and Saito, A. (eds.) 2012. *Locating Neoliberalism in East Asia: Neo- Liberalizing Spaces in Developmental States*. Chichester: Wiley Blackwell.

Peck, J., Theodore, N. and Brenner, N. 2009. "Neoliberal urbanism: models, moments, mutations." *SAIS Review of International Affairs* 29(1): 49–66.

Pirie, I. 2008. *The Korean Developmental State: From Dirigisme to Neoliberalism*. London: Routledge.

POSCO E&C. 2004. "10-year History of POSCO E&C (1994–2004)." Accessed on 9 September 2015 from http://history.poscoenc.com/housing/housing_01_13.htm.

Segel, A.I. 2012. *New Songdo City: HBS Cases 9-206-019* (Revised edition). Boston, MA: Harvard Business School.

Shatkin, G. 2011. "Planning privatopolis: representation and contestation in the development of urban integrated mega-projects." In Roy, A. and Ong, A. (eds.) *Worlding Cities: Asian Experiments and the Art of Being Global*. London: Blackwell, 77–97.

Shin, H.B. 2009. "Residential redevelopment and entrepreneurial local state: the implication of Beijing's shifting emphasis on urban redevelopment policies." *Urban Studies* 46(13): 2815–2839.

——. 2011. "Vertical accumulation and accelerated urbanism: the East Asian experience." In Gandy, M. (ed.) *Urban Constellations*. Berlin: Jovis Publishers, 48–53.

——. 2014. "Contesting speculative urbanisation and strategising discontents." *City: Analysis of Urban Trends, Culture, Theory, Policy, Action* 18(4–5): 509–516.

——. 2016a. "China meets Korea: the Asian Games, entrepreneurial local states and debt-driven development." In Gruneau, R. and Horne, J. (eds.) *Mega Events and Globalization: Capital and Spectacle in a Changing World Order*. London: Routledge, 186–205.

——. 2016b. "Economic transition and speculative urbanisation in China: gentrification versus dispossession." *Urban Studies* 53(3): 471–489.

Shin, H.B. and Kim, S-H. 2016. "The developmental state, speculative urbanisation and the politics of displacement in gentrifying Seoul." *Urban Studies* 53(3): 540–559.

Shin, H. Park, S.H. and Sonn, J.W. Forthcoming. "The emergence of a multi-scalar growth regime and scalar tension: the politics of urban development in Songdo New City, South Korea." *Environment and Planning C*.

Shin, J-S. (ed.) 2007. *Global Challenges and Local Responses*. London: Routledge.

Shwayri, S.T. 2013. "A model Korean ubiquitous eco-city? The politics of making Songdo." *Journal of Urban Technology* 20(1): 39–55.

Song, J-A. 2004. "S Korea sees new life in city rising from the mud." *Financial Times*, 18 August. Accessed on 9 September 2015 from http://on.ft.com/1ydP5nz.

The Dong-A Ilbo. 1995. "Special press interview with three mayoral candidates in Incheon for 27 June election (6.27 seongeo teugbyeol hoigyeon Incheon sijang hubo 3 in)." 4 June. 5 [in Korean].

——. 1997. "Incheon Songdo media valley dreams to become the 'mecca of information and communication industry' (Incheon Songdo midieo baelli 'jeongbo tongsin saneop meka' ggumggunda)." 8 December. 35 [in Korean].

——. 2000. "Difficulties in constructing Songdo City (Songdo sindosi geonseol nanhang". 12 June. 27 [in Korean].

The Hankyoreh. 1993. "Songdo new marine city plan seeing six years wasted (Songdo haesang sindosi gyehoig cheot sabjil handamyeo 6 nyeon heulleo)." 13 August. 12 [in Korean].

——. 1994. "Incheon Songdo seeing the ground breaking of a new marine city (Incheon Songdo e haesang sindosi chaggong)." 11 September. 2 [in Korean].

——. 1998. "A fiasco in the construction of Songdo New City (Songdo sae dosi joseong saeop chajil)." 11 February. 25 [in Korean].

The KyungHyang Shinmun. 1962. "A new earth (Ddo hanaeui daeji)." 10 January. 2 [in Korean].

Woo-Cumings, M. (ed.) 1999. *The Developmental State.* Ithaca, NY; London: Cornell University Press.

Wu, F. and Phelps, N. 2011. "(Post)suburban development and state entrepreneurialism in Beijing's outer suburbs." *Environment and Planning A* 43: 410–430.

Yigitcanlar, T. and Lee, S.H. 2014. "Korean ubiquitous-eco-city: a smart sustainable urban form or a branding hoax." *Technological Forecasting and Social Change* 89: 100–114.

6

FROM PETRO-URBANISM
TO KNOWLEDGE MEGAPROJECTS
IN THE PERSIAN GULF

Qatar Foundation's Education City

Agatino Rizzo

Introduction

Arab countries of the Persian Gulf Region are emerging as important economic and geopolitical hubs. Their substantial oil and natural gas surpluses make them important actors and partners to reboot troubled Western economies (Seznec 2008, Bahgat 2009, Spencer and Kinninmont 2013, 49). "[Estimates] from the consulting firm McKinsey show an accumulation of oil income in the Gulf Cooperation Council (GCC) countries at $2.4 trillion by 2010 and $8.8 trillion by 2020" (Seznec, 2008: 97). It is estimated that a large share of this wealth will be invested outside the GCC "and where better than in shares in the major financial institutions of Wall Street, which have suffered seriously from their less-than-wise investments in subprime mortgage paper"? (ibid. 97.)

Moreover, due to their geographical position and large financial resources (the result of oil and gas revenues), Gulf countries have also managed to become important global transportation hubs for international passengers between the West and Asia and beyond it (Henderson 2006, O'Connell 2006). "This region is also leading the world in aircraft orders as $60 billion has been invested by just three airlines with $27 billion ordered in 2005 alone" (O'Connell 2006, 94). Tourism is one of the pillars of some regional economies such as Dubai's. This latter city has *de facto* set the standards for economic diversification for other neighbouring resource-dependent economies (Henderson 2006, 91).

At the same time, Arab Gulf countries, such as the small emirate of Qatar, have been keen to redraw the future of post Arab-Spring countries in the Middle East and to further geopolitical cooperation between the Gulf and the West. To this end, the creation of Arab-based satellite news channels such as Qatar's Al Jazeera has contributed to extend the sphere of influence of once geopolitically, peripheral Gulf countries (Seib 2008). Philip Seib has labelled this phenomenon the "Al Jazeera Effect" – a term that in his view "encompasses the use of new media as tools in every aspect of the global affairs, ranging from democratization to terrorism" (Seib 2008, IX). However, Rasha El-Ibiary (2011, 200) explains better that the "Al Jazeera Effect" is all but a push for democracy in the Gulf because first and foremost new, Arab-based, all-news channels fail to scrutinize Gulf countries' home-policy – these latter characterized by the lack of basic civil rights (such as freedom of expression). For her (ibid. 202), Al Jazeera is nothing more than a tool to protect Qatar from Saudi influence and to project Qatar onto the world stage. Therefore, "perhaps counter-intuitively, the Arab uprisings have so far resulted in the Gulf Cooperation Council and Western countries working more closely together" (Spencer and Kinninmont 2013, 49) rather than re-composing disagreements within GCC members (e.g., between Qatar and Saudi Arabia[1]).

In the last 15 years, all the factors highlighted above have contributed to re-shape the economic profile of Gulf countries and to fuel rapid demographic, economic, and urban growth in their capitals. In particular in Doha, the capital of Qatar, between 2004 and 2010 residents have more than doubled – from 400 to nearly 800 thousand inhabitants - while, according to the World Bank database (2011), in 2011 alone Qatar GDP grew by 18.8 per cent. To cope with this extraordinary growth, and similarly to other neighbouring Arab capitals (e.g. Dubai, Abu Dhabi, Manama), Qatar has embarked on a massive modernization of its urban infrastructure through government-led urban megaprojects (Rizzo 2014).

Megaprojects are not a new phenomenon. Under colonial rules, the construction of large infrastructures such as railways, dams, canals, and harbours to facilitate the exploitation of local resources was a typical phenomenon. Between the 1950s and 1970s, during the "great mega-project era" (Altshuler and Luberoff 2003), state-initiated megaprojects helped to reconstruct post-WWII Europe (Orueta and Fainstein 2008, 759) and to further the "ideal of democratizing society and distributing a 'fair share' of their benefits" (Lehrer and Laidley 2008, 788). Gellert and Lynch (2003, 16) differentiate between four types of megaprojects: *infrastructure* (e.g., dams, railways, etc.); *extraction* (e.g., minerals, oil and gas); *production* (industrial parks, economic zones, etc.); and *consumption* (e.g., theme parks, real estate developments, etc.).

New megaprojects are implemented as collaboration between state and private actors (Lehrer and Laidley 2008, 789). In the autocratic monarchies of the Gulf, consumption megaprojects are large-scale, government-backed, urban interventions that officially attempt to both modernize and improve the competitive edge of their capitals in the context of economic globalization. One of the main features of current consumption megaprojects in Asia is the ability to "transform landscapes rapidly, intentionally, and profoundly in very visible ways" (Gellert and Lynch 2003, 15).

However, countries such as Qatar are also investing in innovative, integrated developments which I have labelled *Knowledge Megaprojects*. In Knowledge Megaprojects the political and financial commitment from national governments is higher to compensate for the loss of consumption-oriented activities. Knowledge Megaprojects are clearly a typology apart from those analysed by Orueta and Fainstein (2008, 760) – whereby the State is more an instigator and facilitator of big projects rather than the main shareholder. In Doha, for example, Qatar Foundation, a not-for-profit governmental organization that aims to facilitate Qatar transition from a carbon-driven to a knowledge-intensive economy, is building a brand-new university campus (Education City) for foreign universities and a science and technology park that when completed will also include a major specialized teaching hospital, central library, convention centre, equestrian academy, golf club, five-star hotels, and large-scale on-campus housing.

In this chapter, after reviewing the intertwined effects of geopolitical aspirations and the rapid urbanization of Gulf capitals, I will focus on Qatar Foundation's Education City campus to analyse processes and politics of Knowledge Megaprojects in resource-rich countries of the Arab Gulf Region. I argue that while Education City is a new type of Knowledge Megaproject, in that it aims to be a knowledge hub that is better connected with other urban localities around the world to produce knowledge and innovation (Ascher 1995, Castells 1996), its outcomes are similar to those of other consumption megaprojects in the Gulf, i.e. social displacement, spatial segregation, and the privatization of urban space (Graham and Marvin 2001, Gellert and Lynch 2003, Bagaeen 2007).

Globalization and geopolitical aspirations in Qatar

In recent years the small emirate of Qatar in the Arab Gulf region has gained the attention of the media and the international community for its unprecedented economic performance and rising geopolitical ambitions. According to World Bank statistics, the Arab Gulf Region is one of the fastest growing economies in the world. In 2008, while Western economies struggled to deal with the first impacts of the Lehman Brothers collapse, the Gross Domestic Product (GDP) growth rate of the UAE, Bahrain, and Qatar was respectively 5.1 per cent, 6.3 per cent, and 25 per cent (World Bank 2011). Moreover, in recent years, thanks to increasing oil and gas prices, GDP per capita in Qatar has risen by 9 per cent, making Qatar one of the richest countries in the world. As result of this growth and a wider global economic restructuring, a growing number of international companies have moved their regional headquarters to Qatar to benefit from the country's tremendous economic growth.

However, international comparative studies on global cities and networks underestimate the importance of the Gulf Region for contemporary city making. In Peter Taylor's recent Global Urban Analysis (2011), the city of Doha is positioned only in 12th place among 17 cities of the Middle East and North Africa (MENA) region for global network connectivity.[2] When compared to London (at the top of a list of 575 cities worldwide – Taylor et al. 2011, 25), Doha has only 21 per cent (with London = 100 per cent) of the British capital's service, media, and financial firms' global connectivity (Bassens et al. 2011a, 104), while in the same chart Istanbul, Tel Aviv, and Dubai score respectively 53 per cent, 41 per cent, and 40 per cent. More specifically, when analysing the results in the smaller context of the Arab Gulf Region (Bassens et al. 2011b, 284), Doha's global connectivity (21 per cent) is outdone by respectively Dubai (40 per cent), Riyadh (37 per cent), Kuwait City (35 per cent), and Manama (26 per cent), while Muscat (12 per cent) and Abu Dhabi (10 per cent) score less than Doha. According to Bassens et al. (2011b, 284–286), while Dubai is becoming an "all-round service centre" and Manama is an important regional hub, Doha's economy still caters for the local market rather than for the global one. Doha's lower position in the rank reflects other cities' success "in becoming important nodes of these networks of advanced servicing" (Bassens et al. 2011, 283).

The above analysis contrasts with the increasingly important role of Qatar for the global economy and the international community that is the basis for Gulf-based scholars' studies of the region (see, among others, the work of Elsheshtawy 2004 and 2008, Nagy 2006, Pacione 2005, Ponzini 2011, Rizzo 2013, Rizzo 2016). In the case of Qatar, but this can be generalized to include other Arab Gulf countries, on the footprints of long-established Sovereign Wealth Funds (SWFs) in the region (Davidson 2009, 62), Qatar's SWF (managed by the Qatar Investment Authority or QIA) has acquired several stakes in European banks (notably Barclays and Credit Suisse) and automobile industries (Volkswagen and Porsche) while it has been actively investing in real estate again in Europe and in Africa through its controlled Qatari Diar (the premier real estate company owned by the State of Qatar). Yet in the strategic aviation sector, the government of Qatar directly controls Qatar Airways, this latter being one of the biggest customers for Airbus (*The Economist* 2010, Haberly 2011, 1839), and currently owns 20% of IAG, the parent company for British Airways, Iberia, Aer Lingus, and Vueling. This means that Qatar sits on the boards of the major Western corporations and banks that are politically and economically strategic worldwide. This sort of *reverse colonization* of Western assets is not uncommon for Asian emerging economies such as Saudi Arabia, Abu Dhabi, and Malaysia. This trend has been suspiciously scrutinized by Western countries, afraid of being manipulated by non-transparent, government-owned, foreign funds (Bahgat 2009, 1201). As a result of this rising concern, SWFs have been pressured to agree on a set of principles (IMF's Santiago Declaration) that should guarantee their transparency (ibid. 1203).

Second, Qatar is leveraging on its state-owned[3] Al Jazeera satellite news-channel to extend its geopolitical sphere of influence into other Middle East countries (*Financial Times* 2011). With its coverage of the Arab revolution, Al Jazeera has connected the stories of activists in Tunisia with those in Egypt, Libya, Syria, and Yemen in an effort to force out long-running autocratic regimes. Moreover, while peaceful relations with both Iran and Saudi Arabia secure its political stability and wealth, Qatar was one of the few Arab countries actively participating to the NATO strikes in Libya and is presently demanding military intervention in Syria. In addition to this, Qatar is an active player in facilitating, among others, Taliban-NATO discussions over an agreement to end the Afghan conflict (*The Economist* 2012) and the ideological convergence of Syrian opposition parties to end the Bashar Al Assad regime (BBC News 2012). As explained in the introductory section, Qatar's geopolitical positions aim, on the one hand, to protect its regional interests against those of the biggest regional powers such as Saudi Arabia and Iran (El-Ibiary 2011, 201); on the other hand, I argue Qatar's involvement in international politics is the direct result of its expansionist economic agenda. That is, for Qatar to be able to operate in Western markets (e.g., purchase of strategic assets in Deutsche Bank, Credit Swiss, Total, and other French military contractors), full support of Western (i.e. US and EU) foreign policy is required.

Finally, Qatar will be the host of several global events such as the prestigious 2022 FIFA World Cup. Sports events are particularly important to put the country on the map and to increase the number of tourists visiting the country. Yet sport plays an important role in fulfilling the Qatar National Vision 2030[4] target to ensure a healthy lifestyle for Qataris. In the last 10 years Qatar's natural resource-driven wealth has supported a number of events including the 2006 Asian Olympic Games, the 2011 Pan Arab Games, the 2011 Asian cup, and yearly events including, among others, the Qatar Motorcycle Grand Prix, the WTA and ATP Tennis Tour Championships, and the PGA Golf European Tour. Qatar has also put forward bids for the 2016 Summer Olympics, the 2017 World Championships in Athletics, and the 2020 Summer Olympics. This mammoth effort has resulted in massive government-sponsored investments in infrastructures across Doha such as new airports, ports, highways, football stadia, swimming complexes, hotels, and so on (see Figure 6.1).

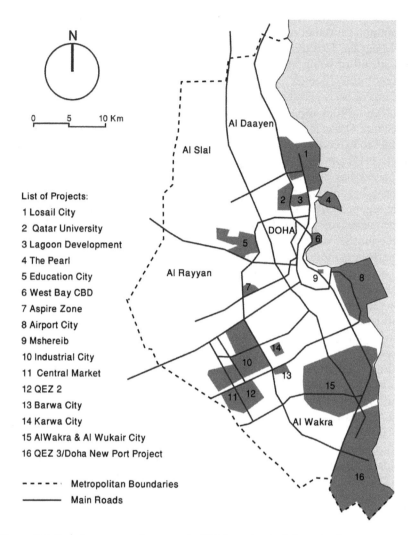

N

0   5   10 Km

Al Daayen

Al Slal

List of Projects:
1 Losail City
2 Qatar University
3 Lagoon Development
4 The Pearl
5 Education City
6 West Bay CBD
7 Aspire Zone
8 Airport City
9 Mshereib
10 Industrial City
11 Central Market
12 QEZ 2
13 Barwa City
14 Karwa City
15 AlWakra & Al Wukair City
16 QEZ 3/Doha New Port Project

- - - - -   Metropolitan Boundaries
————   Main Roads

DOHA

Al Rayyan

Al Wakra

Figure 6.1 Doha's mega-project agenda (2013, author's elaboration)

At the same time, the Gulf is a location for a very diverse population made of temporary workers (Nagy 2006) which, in the case of Qatar and the UAE, outnumber 5 to 1 nationals. According to recent statistics (US State Department 2008), Qatari citizens account for 20 per cent of the total population while South and South East Asians account for 55 per cent. In 2009 in UAE, these proportions are respectively 15 per cent and 67 per cent (US State Department 2014). Also, due to foreign immigration, between 2004 and 2010 Metro Doha's population has more than

106

doubled from nearly 700,000 to 1.4 million inhabitants while the Ministry of Municipality and Urban Planning (MMUP) forecasts that by 2032 the Metro Doha population will be 1.8 million inhabitants (Rizzo 2013). In the last 15 years, Doha has grown from a small, port city to a bustling capital region with global ambitions by increasing its pre-independence population by 30 times. Today almost 90 per cent of Qatar's total population (1.4 out of 1.7 million inhabitants) live in metropolitan Doha. The population growth between the two last general surveys (in 2004 and 2010) was 128 per cent, at the country level, and 93 per cent in Doha. In the same period, in other Gulf emirates, the population grew 11 per cent in Abu Dhabi, 21 per cent in Dubai, and 16 per cent in the whole UAE.

Since foreign workers are offered salaries that are comparable to those of their country of origin, rather than being based on the actual level of experience (Nagy 2006, 124), ethnic diversity and income inequalities are very much correlated in the Gulf economies.[5] Unfortunately, to date, data concerning income by nationality is not available. However, empirical observations suggest that poorly educated and non-Western-educated immigrants have generally the lowest salary while Qatari and Western-educated professionals have the highest (Rizzo 2013, 534). The *Financial Times* (2011) reports that the average basic salary for an immigrant worker in Qatar is approximately QAR 600 (US$165) a month, plus a QAR 200 food allowance, while at the same time immigrants are required to pay more than QAR 4,000 to immigrate to Qatar (agency expenses, trips, and VISA). These inequalities are reinforced by an "archaic sponsorship system in which workers are forbidden from leaving a company without its permission" (*Financial Times* 2011).

Qatar's diverse population, however, is never taken into account by local planners and decision makers. Elsheshtawy (2004, 172) notes that despite "a happy merging of cultures . . . no real effort is made [in Dubai] to resolve social problems, address concerns of the [expatriate] lower class, or try to make the urban environment more 'livable'". While by 2022 more luxury hotels will be opened in Doha to cater for Westerns and westernized Arabs who can afford to enjoy exclusive services including spa treatments, gourmet restaurants, and fancy bars or pubs serving alcoholic drinks, issues such as spatial segregation, gentrification of redeveloped areas, and affordable housing remain still unsolved (Rizzo 2013).

While this megaprojects agenda has brought new useful infrastructures to cater for the country's ambitious goals, at the same time, due to the poor planning regulations (Rizzo 2014, 51) and in absence of democratic processes (Ponzini 2011), it has been responsible for important physical and social fractures within the city – i.e. between who can and who cannot access those infrastructures (Rizzo 2013). These latter aspects will be further discussed in the next section.

Petro-urbanism and petropolitics

The Gulf economy was previously based on daily fishing, seasonal pearling, and sea-borne trade with other countries facing the Persian Gulf (e.g., Iran, Trucial States, etc.). The exploitation of oil from the 1950s altered this profile in favour of oil-related activities and public-sector jobs (Riad 1981). The oil industry has brought a dramatic change that has altered for ever the natural (marine and desert ecosystems), urban (cities and infrastructures) and socio-economic (consumption patterns) balance of Gulf countries. However, it is worth noting that this sudden change has not managed to re-shape with the same pace the socio-religious values of the Gulf (e.g., status of women, patriarchal hierarchy, etc. – Riad 1981, 8 and 10) towards Western ones (e.g., democracy, human rights, and civil society).

While international literature on Qatar and the Gulf Region is still in the making (see recent attempts by Koolhaas et al. 2007 and 2010 with Al Manakh and Al Manakh Gulf Continued), a number of academic essays and articles published by Gulf-based researchers (Riad 1981, Elsheshtawy 2004 and 2008, Nagy 2006, Pacione 2005, Ponzini 2011, Rizzo 2013, Rizzo 2014) have attempted to sketch the dynamics and impacts of rapid urbanization in the Gulf Region.

Several terms have been used to label the recent extraordinary urban growth of Arab capitals of the Persian Gulf. Bagaeen (2007, 174) described Gulf cities with the term "instant urbanism" to differentiate them from Western, long-term urban evolution. This is for instance the case of Dubai, Abu Dhabi, and Doha which have all transformed from small port cities to sprawling metropolitan areas in just two decades (Pacione 2005, Ponzini 2011, Rizzo 2013). The urban pattern and demographics of the Gulf have gone through dramatic changes from scattered small settlements towards ever larger "virtual city-states" (Riad 1981, 20). The size of the typical Gulf settlement was previously influenced by the amount of fresh water available, geographical location in respect to pearling grounds, accessibility to inland trade routes towards other settlements of the Arabian peninsula, and political/tribal alliances to secure stable trade. With the advent of the oil industry in the 1950s, these rules were broken for ever and the size of settlements was determined by the availability of public jobs and modern infrastructures (e.g., electricity, water supply, sanitation, etc.). Riad (1981, 7) has labelled this transformation with the term "petro-urbanism" which, according to him, "undermined, with unparalleled suddenness, the roots of an ecosystem [the Gulf's one] which reflected a perfect adaptation to an environment many generations old".

Recently, the term "Dubaization" or "Dubaification" (Elsheshtawy 2010, 250) has come to the foreground of academic literature to describe cities' emulation of Dubai's urban megaprojects agenda – this latter being a pervasive process that, as shown by Choplin and Franck (2010) for Khartoum, Sudan and Nouakchott, Mauritania and by Haines (2011, 160), is not restricted to

Gulf capitals only. In Qatar, for instance, "The Pearl," a gigantic oyster-like human-made island with luxurious residential areas inspired by Venice and Southern France's Riviera, is reminiscent of Dubai's "Palms".

From a geopolitical point of view, Gulf Cities deploy the "the political and economic engine of urban development projects" to compete for Foreign Direct Investment (FDI), skilled labour, and international tourists (Ponzini 2011, 251). A typical Gulf Megaproject is a government-funded, large, integrated urban development characterized by a specific theme (e.g., ancient Arab City, Global City, Mediterranean Riviera City, etc.) and a dominant function (e.g., sport, business, residential, etc.). Literature on Asian megaprojects show that, "developers began to realize that great opportunity lay in building integrated megaprojects" (Shatkin 2008, 391) since each function (residential, office, etc.) feeds into the other, thus amplifying market profitability. Successful megaproject developments in cities such as Singapore have represented for oil-rich Arab rulers a model to diversify their economies and modernize their capital cities. However, in the case of Gulf cities, finance for these projects has been largely provided by the state through oil revenues. In the case of oil-poor Dubai, State's finances have been integrated with funds from resource-rich neighbours (mainly Abu Dhabi and Saudi Arabia – Bloch 2013).

However, as Ponzini (2011, 252) has pointed out for Abu Dhabi, large-scale projects are neither a new formula nor immune to uncertainties and unbalanced effects – e.g. in 2009 Burji Khalifa in Dubai had to be bailed out with Abu Dhabi's money.[6] Megaprojects in the Gulf Region see the participation of a limited number of actors – usually the government's linked companies – and their implications are never discussed with the public. In Qatar, just like in Abu Dhabi, "the separation between public and private sectors . . . is practically non-existent because the actors have key positions in public decision making and in the management of private companies" (Ponzini 2011, 254). This situation prevents the participation of the inhabitants in decision making and, I argue, furthers the rift between governments and their citizens and between them and the immigrant communities. At the moment these fractures are filled with petrodollars to buy political consensus of nationals – i.e. through well paid government jobs and subsidies – and strict visa-sponsorship regulations to keep away unwanted immigrants. According to Friedman (2006, 28), "[t]he price of oil and the pace of freedom always move in opposite directions in oil-rich petrolist states". This "First Law of Petropolitics" implies that "the pace of freedom really starts to decline as the price of oil really starts to take off" (ibid.). The recent uprisings in countries that have recently run out of oil such as Bahrain suggest that the current economic contract between government and people will survive as long as oil and gas wells will be running. Beyond that point, I argue, a renegotiation of power between the ruling elite and the rest of the inhabitants will be unavoidable.

In a recent special issue on Qatar, the *Financial Times* reported regional analysts' worries over the government's lack of capacity to deliver the massive developments it plans to undertake. "Qatar is enormously wealthy but it doesn't really have the wherewithal in government yet to manage the process in ways that are directly linked to its own ambitions . . . without the software, the hardware is not going to work properly" (*Financial Times* 2011, 4). Little coordination of public and private agencies in the absence of a comprehensive urban master plan is, in fact, one of the main problems of Qatar (Rizzo, 2014). For instance, while in the country there is an oversupply of hotel rooms and luxury residences, government-planned, high-end projects continue to be delivered (though at a slower pace) – e.g. while Qatari Diar is working on the 38-square-kilometre Lusail City for 450,000 inhabitants and workers, Mshereib Properties (a subsidiary of QF) is about to deliver a 35-hectare redevelopment project in the central district of Mshereib, and UDC (a private consortium recently bailed out by the government) has already completed 50 per cent of The Pearl – a 400-hectare human-made island for 41,000 "international residents" (The Pearl Qatar 2013).

Thus in the absence of democratic participation and with little coordination between public agencies, megaprojects in the Gulf are in many cases disconnected with the surrounding built environment. In Qatar, for instance, by analysing Doha's satellite pictures over a long period of time, El Samahy and Hutzel (2010, 187) detected a prevalent pattern of disconnected urban/non-urban areas throughout the Doha metropolitan area. In the next section, I shall discuss Qatar Foundation's Education City, a R&D plus higher education campus located to the West of Doha. Unlike other consumption megaprojects, Education City aims to build capacity among local graduates to steer knowledge-intensive business in the country (this latter being one of the strategic pillars for economic diversification underpinned in the Qatar National Vision 2030). Knowledge Megaprojects are increasingly becoming popular in the Asian emerging markets (Altbach and Knight 2007, Cao 2011). In the Gulf, Dubai's Internet City (opened in 2000) and Media City (opened in 2001), Abu Dhabi's Masdar City and Saudi Arabia's King Abdullah Economic City (both currently under development) show the importance that such knowledge megaprojects have acquired in resource-dependent economies.

Transitioning to a knowledge-based economy: Qatar Foundation's Education City

In the past two decades, in emerging economies such as China, Malaysia, Singapore, and UAE, scholars have reported the rising of the *branch campus* phenomenon (Cao 2011, 8). Prestigious foreign universities are opening branch faculties abroad to tap the internationalization of the higher education market in Asia (Altbach and Knight 2007, 290–291). These campuses are a new form of post-industrial urbanism capable of producing a "milieu of innovation [that is] aimed at generating new knowledge, new processes, and new products" (Castells 1996, 419). In a way, international branch campuses are the manifestation of Castells' and Halls' (1994) Technopoles. However, as I shall illustrate in the next sub-sections, branch campuses in the Gulf are rather complex, integrated, and themed *Knowledge Megaprojects*, characterized by specific intra-national politics and processes.

Although Gulf countries (with the partial exception of Saudi Arabia) have failed to become industrial economies (Ewers and Malecki 2010), on the other hand, they have started to realize the importance of economic diversification towards the post-carbon economy. In particular, Ewers and Malecki (2010, 495) have suggested that importing foreign, high-skilled labour to Gulf countries is one of the main strategies to "leapfrog" the current, resource-based economy. The transition to the knowledge economy is achieved not only by offering competitive financial (i.e. attractive, tax-free salaries) and non-financial (e.g., free housing, subsidised health care, return tickets to home countries for summer holidays, etc.) benefits to the ambitious knowledge workers but also by implementing state-of-the-art infrastructures (e.g., Education City campuses) that may facilitate this transition.

While in countries such as Singapore the branch campus phenomenon is reaching saturation, in other countries it is forecasted to rise (Cao 2011, 9). This is particularly true in the Gulf Region which is at the same time home to one of the youngest population of the planet and in need of quality higher education (Healey 2008). While after 9/11 events, strict visa regulations in Western countries (i.e. the US and UK) have limited traditional student/academic mobility from Asia (Koolhaas 2007, 324, Sirat 2008, 88), shrinking resources for higher education and ageing population have persuaded universities in the West to become mobile – to seek bigger, growing markets in the East.

From the New York Institute of Technology in Bahrain to the Box Hill Institute and the Kuwait Maastricht Business School in Kuwait, to the Carnegie Mellon University, Virginia Commonwealth University and several other US universities in Qatar, to the University of Wollongong, Murdoch University, Heriot-Watt University and Middlesex University in Dubai, Wilkins (2011, 74–76) notes that the Gulf States have been the "largest recipients of transnational higher education globally, whilst Australia, the UK and USA have been the largest providers". In many cases foreign universities have been beneficial to the development of the local economy by providing skills in strategic sectors such as banking, engineering, and so forth. These latter skills are crucial to facilitate the diversification of the Gulf economies away from the oil and gas dominance (Peterson 2009, 11, Davidson 2009, 65). The internationalization of higher education can also bring savings to national governments, "in that they do not have to bear the cost of those students' education, as they do not usually pay the tuition fees or any of the associated costs of study" (ibid. 77–78).

However, although the implementation of Knowledge Megaprojects to transition to a cleaner economy is a sound and noble aim, the processes and politics behind this are not so different from those characterizing other consumption megaprojects in the Gulf (e.g., Dubai Marina, Doha's The Pearl, etc.). In the remainder of this section I will unravel the main driving forces behind the Education City to exemplify the intertwined effects of pressures for economic modernization, rapid urbanization, and undemocratic governance.

Education City: an overview of the knowledge megaproject

Qatar Foundation has inspired a new generation of leaders. In a short time I have seen our graduates taking serious positions in various ministries and corporations here in Qatar, starting up their own successful business ventures, and undertaking scientific research. This is just the beginning.[7]

From 1996 Qatar Foundation for Education, Science and Community Development (QF), a non-profit governmental organization funded mainly with oil revenues, has worked on the promotion and implementation of an education and science & technology campus in Doha to facilitate Qatar's transition to the knowledge-based economy (Rizzo 2013). Education City is a 14-square-kilometre area which when completed will include, among others, 15 American and European universities, a major specialized teaching hospital, central library, convention centre, equestrian academy, 18-hole golf club, large-scale on-campus housing, and a science & technology park (see Figure 6.2). QF has invited architects from all over the world to work on this ambitious knowledge megaproject (Rizzo 2013).

Texas A&M University, Weill Cornell Medical College, Georgetown University, Virginia Commonwealth University, Carnegie Mellon University, Northwestern University, HEC Paris, and University College London have already based their activities in Education City. The science & technology park, also located in Education City area, is working with dozens of multinational corporations including Exxon-Mobil, Cisco, and Microsoft to deliver Qatar's ambitions to be a knowledge-intensive hub.

Figure 6.2 Qatar Foundation's Education City: to the North the Science &
Technology Park; International Universities are located at the centre of
the development; to the South the stadium and sport facilities

Source: Google Earth, 2014

Education City: space of flows

Between February and March 2012, I carried out a series of interviews with
QF employees in the American universities and private research clusters of
the science & technology park with the aim of studying knowledge megaprojects in the Gulf. With a qualitative/explorative approach I asked them questions to understand the quantity and quality of interactions between the
branches and their home campuses abroad. Also, I tried to explore whether
there are periodical flows of visiting staff, faculty, researchers, and/or students between the Qatar branches and the home campuses and vice versa.
Furthermore, when discussing career development for employees of branch
campuses, I inquired about the type of activities (training, meetings, etc.) my
participants were involved in when travelling back to their home campuses.
For the sake of space, I report only excerpts of one interview held with a
participant from the Weill Cornell Medical College in Qatar.

The results of my interviews show how determinant are day-to-day, material, and immaterial flows of communication between the faculty branch

113

and home campus (*I deal almost on a daily basis with the home campuses either in New York City or in Ithaca, NY*). Immaterial flows between Qatar's branch and its home campus in the US are facilitated mainly by normal emails as well as other forms of remote communication (*Our communication is primarily through emails . . . but my boss also periodically talks directly via phone to the administration departments in the U.S.*). However, as the participant explains, these immaterial flows of information are integrated with periodical material flows of documents, staff, faculty, and students between Qatar and the US (e.g., *we have a Weill Cornell Medical College at Qatar office in NY; all mail that needs to be sent directly e.g. signed original contracts, employment offers, some interlibrary loans coming from U.S. library, insurance claims for employees, etc. are sent daily to this office in New York city (going express via Fed Ex) and staff in this office either take the paperwork personally to administrators/departments in New York or forward them on to departments via interoffice mail*). These two forms (material and immaterial) of communication help to coordinate the activities of the branch faculty with their mother institution. The exchanges of information and people between the branch and home campus are not one-way – the branch has an active role to determine students' study plans and training programmes.

Moreover, by mapping companies' headquarters and universities' home campuses, and in light of the knowledge I gleaned through several other informal conversations with employees of branch campuses and companies at QF, I was able to qualitatively depict the global reach of Education City. As we shall see in the next paragraph, this analysis confirmed a global trend "towards uneven global connection combined with an apparently paradoxical trend towards the reinforcement of local boundaries" (Graham and Marvin 2001, 9). Most agencies are located in North America (particularly in the US) and North Europe (particularly in the UK) while few of them are based in the East or have regional offices in the Gulf (Figures 6.3 and 6.4).

Education City: a knowledge megaproject with the
limitations of a consumption development

While Education City links with foreign spaces are strong (in line with Ascher's (1995) work on Metapolis and Castells' (1996) work on the Space of Flows), on the other hand, the campus shows the tendency to be physically and socially disconnected from its urban surroundings. Leaving aside the environmental aspects related to irrigating the vast lawns that separate academic buildings in an arid country, the campus itself is a gigantic gated community with few, heavy, guarded access gates (Rizzo, 2016). A chaotic landscape of non-urbanized or poor urbanized areas encircles the campus

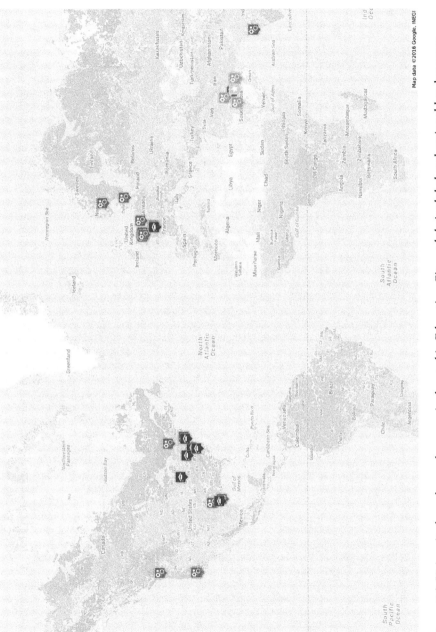

Figure 6.3 University branches and companies located in Education City and their global and regional headquarters

Map data ©2016 Google, INEGI

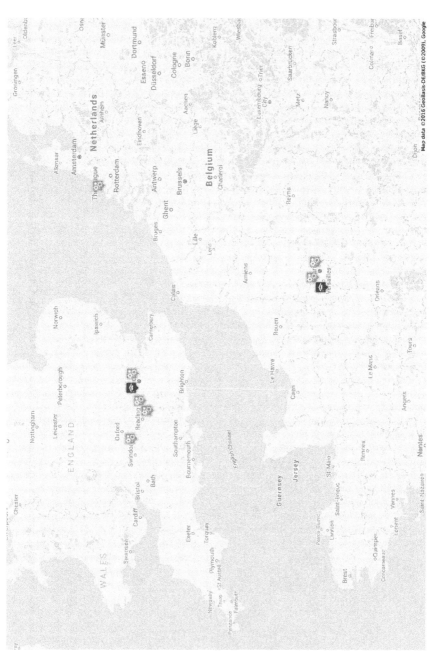

Figure 6.4 University branches and companies located in Education City and their global and regional headquarters: North Europe

while its architecture is visually and typologically alien to that of the sur-rounding city (Rizzo, 2013).

Moreover, Education City reproduces the very same ethnic-based patterns as other consumption megaprojects do, i.e. South Asians relegated to the bottom of the social hierarchy while nationals and Western-educated professionals stay at the top. Gellert and Lynch (2003) have argued that most often megaprojects act as devices for bio-geophysical and social displacement. They distinguish between two types of displacement: "direct" and "indirect". The former is an integral part of the process that is more predictable while the latter is an indirect consequence that is temporally and/or spatially less immediate and subjected to far greater uncertainty. Gellert and Lynch (2003) note that "[direct] displacement not only refers to the movement of people 'out of the way' of project development, but the movement of workers into areas [working sites and labour camps] where the demand for project labor outstrips the local supply".

While Education City aims to strengthen Doha's links with knowl-edge-intensive cities – mainly located in North America and Europe – to promote Qatar's transition to a post-carbon economy, the campus repli-cates the same pitfalls of other consumption megaprojects in the region i.e. social and physical fragmentation of urban space. Following Graham and Marvin's (2001, 222) notion of the "splintering metropolis," these bundled, high-end, mega-urban developments secured from the influences of poorer groups can also be labelled as "secessionary network space". In other words, in Education City "security, urban design, financial, infrastruc-tural and state practices . . . [are deployed to] . . . separate the social and economic lives of the rich from those of the poor" (ibid. 222). However, while in the Western world this secessionary space is produced by the interplay of international real estate capital and planning and design con-sultancies (ibid. 223), in Qatar, as in the rest of the Gulf, it is financed by the State.

As Mendis (2007, 320) observed, Education City is at the same time both a "resource" for the country's needs and a "commodity" to brand Qatar through its distinctive architecture similar to Abu Dhabi's Masdar City and Dubai's Palms (Pacione 2005, 265, Ponzini 2011, 254). Also the processes behind the implementation of this knowledge campus are similar to those of consumption megaprojects: the Al Thani ruling family and their allies occupy key posts on the board of QF and the Ministries involved in the project. QF projects are announced and delivered without discussing their meanings and implications for the Qatari society: while the campus enjoys a luxury equestrian stadium in which horses are looked after in modern hos-pital structures (Gioni 2011, 82), in the nearby Women's Hospitals ladies "compete for the attention of the staff" (*Financial Times* 2011) in a two-decades-old structure.

Conclusions

The cumulative effects of the Gulf's "instant urbanism" and Qatar's geopolitical aspirations have produced ambitious urban transformations in Doha. This has led to the emergence of a new type of mega development which I have called *Knowledge Megaprojects*. Knowledge Megaprojects in the Gulf are stirred, sponsored, and implemented by a majority, government-linked shareholder unlike traditional megaprojects (Oureta and Fainstein 2008) in which the State acts more as an instigator and facilitator for private capitals. Qatar Foundation's Education City is a large knowledge megaproject hosting several foreign universities and international companies, with the aim to facilitate Qatar's transition from the carbon to post-carbon economy.

However, I argue that, knowledge megaprojects in the autocratic monarchies of the Persian Gulf Region are characterized by the very same politics and planning processes of consumption megaprojects, i.e. powerful families, who are at the same time policy makers and business stakeholders, steer government revenues and replace market forces to benefit from the economic transition (Ponzini 2011). In doing so, knowledge megaprojects contribute to strengthen the already unsustainable levels of spatial segregation and social displacement in the Gulf (Riad 1981, Elsheshtawy 2004, Adham 2006). Flyvbjerg (2005, 18) brilliantly describes the economic and environmental paradoxes of megaprojects with a Machiavellian formula: "a fantasy world of underestimated costs, overestimated revenues, undervalued environmental impacts and overvalued regional development effects".

While at the moment these issues are managed by a generous redistribution of the country's wealth among nationals (see Friedman's (2006) First Law of Petropolitics), it is in the interest of Gulf countries such as Qatar to rethink their strategies to deliver a more just and sustainable cities.

Notes

1　See the recent unilateral withdrawal of Saudi, Bahraini, and Emirati ambassadors from Qatar over allegations of political interference. Muslim Brothers' leaders are openly supported by Qatar and broadcasted in Al Jazeera programmes, and their speeches often called for democracy in the absolute monarchies of the Gulf region.
2　Global Network Connectivity measures the degree of connection of 525 selected cities across the world considering international firms in the following sectors: finance, accountancy, advertising, law, and management consultancy (Taylor 2011, 7).
3　The government of Qatar owns Al Jazeera satellite channels through Qatar Media Corporation.
4　In February 2012 former Emir of Qatar Sheik Hamad created a national day for sports to be held yearly.
5　For example, two professionals having the same experience and skills, working for the same company, but coming from two different countries (say, India and France) are likely to receive two very different salaries – i.e. the French professional will earn substantially more than his Indian colleague (see Nagy, 2006).

6 In 2008 Emaar, a private conglomerate controlled by the Dubai Government, was
   bailed by Abu Dhabi. In exchange for the help, the government in Dubai decided
   to name the world's tallest tower after the name of Abu Dhabi's ruler Sheik Khalifa
   (the previous name for the building was Burj Dubai).
7 Excerpt from the speech of Sheikha Hind bint Hamad Al Thani, QF Vice
   Chairperson cited in *The Peninsula* newspaper, Thursday 8 March 2012, p. 6.

References

Adham, K. 2008. "Rediscovering the island: Doha's urbanity from pearls to specta-
   cle." In *The Evolving Arab City: Tradition, Modernity and Urban Development*,
   by Y. Elsheshtawy (ed.), 218–256. London: Routledge.
Altbach, P.G. and Knight, J. 2006. *The Internationalization of Higher Education:
   Motivations and Realities.* The NEA 2006 Almanac of Higher Education, 3–10.
Altshuler, A. and Luberoff, D. 2003. *Mega-Projects: The Changing Politics of Urban
   Public Investment.* Washington DC: Brookins Institution Press.
Ascher, F. 1995. *Metapolis ou l'avenir des villes.* Paris: Editions Odile Jacobs.
Bagaeen, S. 2007. "Brand Dubai: the instant city; or the instantly recognizable city."
   International Planning Cities 12(2): 173–197.
Bassens, D., Derrudder, B. and Witlox, F. 2011a. "Middle East/North African cities
   in globalization." In *Global Urban Analysis: A Survey of Cities in Globalization*,
   by P.J. Taylor, P. Ni., B. Derudder, M. Hoyler, J. Huang and F. Witlox (eds.),
   102–113. London: Earthscan.
Bassens, D., Derrudder, B. and Witlox, F. 2011b. "Arabian Gulf cities." In *Global Urban
   Analysis: A Survey of Cities in Globalization*, by P.J. Taylor, P. Ni., B. Derudder, M.
   Hoyler, J. Huang and F. Witlox (eds.) (2011), 284–287. London: Earthscan.
BBC News. 2012. *Syria opposition groups hold crucial Qatar Meeting.* 4 November.
   http://www.bbc.co.uk/news/world-middle-east-20196107.
Bhagat, G. 2008. "Sovereign wealth funds: dangers and opportunities." *International
   Affairs* 12(2): 1189–1204.
Bloch, R. 2010. "Dubai's long goodbye." *International Journal of Urban and
   Regional Research* 34(4): 943–951.
Cao, Y. 2011. "Branch campuses in Asia and the pacific: definitions, challenges and
   strategies." *Comparative and International Higher Education* 3: 8–10.
Castells, M. 1996. *The Rise of Network Society: The Information Age: Economy,
   Society, and Culture* (2010 ed.). Chichester: Wiley-Blackwell.
Castells, M. & Hall, P., (1994). *Technopoles of the World. The making of twenty-
   first-century industrial complexes.* London, New York.
Choplin, A. and Franck, A. 2010. "A glimpse of Dubai in Khartoum and Nouakchott:
   prestige urban projects on the margins of the Arab world." *Built Environment*
   36(2): 192–205.
Cornell, J. 2006. "Medical tourism: sea, sun and surgery." *Tourism Management*
   27(6): 1093–1100.
Davidson, C. 2009. "Abu Dhabi's new economy: oil, investment and domestic development." *Middle East Policy* 16(2): 59–79.
El-libiary, R. 2011. "Questioning the Al-Jazeera effect: analysis of Al-Qaeda's media
   strategy and its relationship with Al-Jazeera." *Global Media and Communications*
   7(3): 199–204.

El Samahy, R. and Hutzel, K. 2010. *Closing the Gap.* Vol. 23, in *Al Manakh Gulf Continued*, by R. Koolhass, T. Reisz, M. Gergawi, B. Mendis and T. Decker (eds.), 184–190. Archis: Amsterdam.

Elsheshtawy, Y. 2004. "Redrawing boundaries: Dubai, an emerging city." In *Planning Middle Eastern Cities: An Urban Kaleidoscope in a Globalizing World*, by Y. Elsheshtawy (ed.), 169–199. London: Routledge.

——. 2008. "Cities of sand and fog: Abu Dhabi's global ambitions." In *The Evolving Arab City: Tradition, Modernity and Urban Development*, by Y. Elsheshtawy, 258–300. London: Routledge.

——. 2010. *Dubai: Behind an Urban Spectacle.* London: Routledge.

Energy Information Administration. 2006. *International Carbon Dioxide Emissions and Carbon Intensity Charts.* http://www.eia.gov/emeu/international/carbondi oxide.html.

Ewers, M.C. and Malecki, E.J. 2010. "Leapfrogging into the knowledge economy: assessing the economic development strategies of the Arab Gulf states." *Trjdschrift voor economische en sociale geografie (Journal of Economic and Social Geography)* 101(5): 494–508.

Financial Times. 2011. "Qatar Special Report." 17 December: 1–6.

Flyvbjerg, B. 2005. "Machiavellian megaprojects." *Antipode* 37(1): 18–22.

Friedman, T.L. 2006. "The first law of petropolitics." *Foreign Policy* 154(3): 28–36.

Gellert, P.K. and Lynch, B.D. 2003. "Mega-projects as displacements." *International Social Science Journal* 55(175): 15–25.

General Secretariat for Development Planning. 2011. *Qatar National Vision 2030.* http://www.gsdp.gov.qa/portal/page/portal/gsdp_en/nds.

Gioni, M. 2011. "A Letter from Doha." *Domus* 953: 82–89.

Graham, S. and Marvin, S. 2001. *Splintering Urbanism, Networked Infrastructures, Technological Mobilities and Urban Condition.* London: Routledge.

Haberly, D. 2011. "Strategic sovereign wealth fund investment and the new alliance capitalism: a network mapping investigation." *Environment and Planning A* 43(8): 1833–1852.

Haines, C. 2011. "Cracks in the façade: landscapes of hope and desire in Dubai." In *Worlding Cities: Asian Experiments and the Art of Being Global*, by A. Roy and A. Ong (eds.), 160–181. Malden: Blackwell Publishing Limited.

Healey, N. 2008. "The changing face of higher education in Asia and what it means for Europe." *20th Annual Conference on Redesigning the Map of European Higher Education.* Antwerp.

Henderson, J.C. 2006. "Tourism in Dubai: overcoming barriers to destination development." *International Journal of Tourism Research* 8(2): 87–99.

Joutsiniemi, A. 2010. "Becoming metapolis: a configurational approach." *DATUTOP Occasional Papers* (School of Architecture, Tampere University of Technology, Occasional Papers), 32.

Koolhaas, R. and De Graaf, R. 2007. *The Future of Knowledge.* Vol. 12, in *Al Manakh*, by R. Koolhas, O. Bouman and M. Wigley (eds.), 324–327. Columbia University GSAAP: Stichting Archis.

Koolhaas, R., Reisz, T., Gergawi, M., Mendis, B. and T. Decker (eds.). 2010. *Al Manakh Gulf Continued.* Vol. 23. Columbia University GSAAP: Stichting Archis.

Koolhaas, R., Bouman, O. and Wigley, M. (eds.). 2007. *Al Manakh.* Vol. 12. Columbia University GSAAP: Stichting Archis.

Lehrer, U. and Laidley, J. 2008. "Old mega-projects newly packaged? Waterfront redevelopment in Toronto." *International Journal of Urban and Journal Research* 32(4): 786–803.

Mendis, B. 2007. *Branding Qatar Education City*. Vol. 12, in *Al Manakh*, by R. Koolhaas, O. Bouman and M. Wigley (eds.), 320–325. Columbia University GSAAP: Stichting Archis.

Nagy, S. 2006. "Making room for migrants, making sense of difference: spatial and ideological expressions of social diversity in urban Qatar." *Urban Studies* 43(1): 119–137.

O'Connell, J.F. 2006. "The changing dynamics of the Arab Gulf based airlines and an investigation into the strategies that are making Emirates into a global challenger." *World Review of Intermodal Transportation Research* 1(1): 94–114.

Orueta, F.D. and Fainstein, S.S. 2008. "The new mega-projects: genesis and impacts." *International Journal of Urban and Regional Research* 1(1): 759–767.

Pacione, M. 2005. "City profile Dubai." *Cities* 22(3): 255–265.

Peterson, J.E. 2009. "Life after oil: economic alternatives for the Arab Gulf states." *Mediterranean Quarterly* 20(3): 1–18.

Ponzini, D. 2011. "Large scale development projects and star architecture in the absence of democratic politics: the case of Abu Dhabi, UAE." *Cities* 28(3): 251–259.

Riad, M. 1981. "Some aspects of petro-urbanism in Arab Gulf states." *Bulletin of the Faculty of Humanities and Social Sciences (Qatar University)*, 7–24.

Rizzo, A. 2013. "Metro Doha." *Cities* 31: 533–543.

——. 2014. "Rapid urban development and national master planning in Arab Gulf countries: Qatar as a case study." *Cities* 39: 50–57.

——. 2016. Sustainable urban development and green megaprojects in the Arab states of the Gulf Region: limitations, covert aims, and unintended outcomes in Doha, Qatar, International Planning Studies, DOI: 10.1080/13563475.2016.1182896

Seib, P. 2008. *The Al Jazeera Effect: How the New Global Media are Reshaping World Politics*. USA: Potomac Books.

Seznec, J.F. 2008. "The Gulf sovereign wealth funds: myths and reality." *Middle East Policy* 15(2): 97.

Shatkin, G. 2008. "The city and the bottom line: urban megaprojects and the privatization of planning in Southeast Asia." *Environment and Planning A* 40(2): 383–401.

Sirat, M. 2008. "The impact of September 11 on international student flow into Malaysia: lessons learned." *The International Journal of Asia Pacific Studies* 40(2): 79–95.

Spencer, C. and Kinninmont, J. 2013. "The Arab Spring: the changing dynamics of West-GCC cooperation." Chap. 2 in *The Uneasy Balance: Potential and Challenges of the West's Relations with the Gulf States*, by R. Alcaro and A. Dessi, 49–69. London: Chatham House, The Royal Institute of International Affairs.

Spiess, A. 2008. "Developing adaptive capacity for responding to environmental change in the Arab Gulf states: uncertainties to linking ecosystem conservation, sustainable development and society in authoritarian rentier economies." *Global and Planetary Change* 64(3): 244–252.

121

Taylor, P.J, Ni, P., Derudder, B., Hoyler, M., Huang, J. and Witlox, F. (eds.). 2011. *Global Urban Analysis: A Survey of Cities in Globalization*. London: Earthscan.

The Economist. 2010. "Aviation in the Gulf: rulers of the new silk road: the ambitions of the three Gulf-based 'super-connecting' airlines are bad news for competitors but good news for passengers." 3 June. http://www.economist.com/node/16271573.

——. 2012. "Afghanistan: dial 1 to speak to the Taliban." 4 January. http://www.economist.com/blogs/banyan/2012/01/afghanistan.

The World Bank. 2011. *Development Indicators*. http://data.worldbank.org/indicator/NY.GDP.MKTP.CD.

US Department of State: Diplomacy in Action. 2014. *Country Database*. http://www.state.gov/p/nea/ci/qa/.

Wilkins, S. 2011. "Who benefits from foreign universities in the Arab Gulf states?" *Australian Universities' Review* 53(1): 73–83.

7

"THEIR HOUSES ON OUR LAND"

Perforations and blockades in the planning
of New Town Rajarhat, Kolkata

Ratoola Kundu

Introduction

Kolkata's north-eastern fringe has been steadily changing in the past
decade. Intricate networks of villages, agricultural fields and water bodies have given way to dense, concrete high-rise apartments, steel and glass
malls, large offices, convention halls, international and local hotel chains
connected through concrete roads—all in different stages of construction.
Now and then, there are a few village pockets, temples, mosques, cultivated
farm lands and herds of cows that puncture the glossy new urban landscape. White-collar workers sip cups of tea at the roadside shacks manned
by erstwhile cultivators. This is New Town, Rajarhat—one of the most visible spatial manifestations of economic liberalization and reforms undertaken by the erstwhile Left Front Government (LFG) in West Bengal in the
1990s.[1] Rajarhat New Town, spread over 37 square kilometres and billed
by the state as West Bengal's first "green, eco-friendly, self-sufficient, and
smart city" (Chakraborti 2014) with a projected population of a million
people, reflects the changing trajectory of urban development. This chapter argues that the shift towards a market-based urban development model
(Bose 2012, Roy 2003) has significantly reconfigured political ties and social
relations across the urban periphery in violent, contested and intractable
ways. Politics around land possession and dispossession is central to these
contestations and fleeting coalitions.

The reordering of spaces, lives and livelihoods from rural to urban in
the fringes of Kolkata is embedded in the larger story of post-1991 political
economic transformation in the country. New economic policies, structural reforms and rescaling of power equations between central, state and
local governments centred on cities for economic development have created multiple demands on land as existing and new actors have emerged to
wrest control over the use, exchange, conversion and development of land

(Shatkin 2014). Large-scale urban projects such as new townships, urban corridors, technology parks and Special Economic Zones (SEZs) are instruments of this new spatial order, actively promoted by state governments to woo private investments and global capital. This has been facilitated through a calculated deregulation and suspension of laws through exceptions and exemptions to planning rules (Roy 2010), ostensibly to speed up the process of urbanization. These spatial instruments are strategically posed as "quick-fix" solutions to the complex and emerging needs of urbanization—for new office spaces, housing for the growing aspirational middle class and spaces that can offer an attractive counterpoint to the congested, impoverished, chaotic, unpredictable and deteriorating old cities.

The volatility of peripheral areas of urban centres stems from multiple claims and demands. Peripheral land is available in large swathes at relatively cheaper prices. Compared to the more heavily built-up core city locations, the regulatory frameworks governing land use, change and conversion are more flexible and open to negotiations in the urban periphery (World Bank 2013). Manufacturers and real estate developers in India are increasingly favouring selective peripheral locations over central city locations, where infrastructure is outmoded and building activity is constrained by restrictive development control regulations (World Bank 2013). The spatial and social transformation of urban peripheries is also related to the expansion of the globally connected emerging knowledge economy in India (Benjamin et al. 2008, Goldman 2011)—visible in the IT parks dotting the outskirts of major cities such as Delhi, Bangalore and Kolkata. New, gated and exclusive urban spaces are being created and marketed for an increasingly mobile and aspirational urban middle class (Bose 2012), bypassing the squalor and congestion of the existing megacities and the claims of the urban poor (Bhattacharya and Sanyal 2011). In addition, scholars have argued that emergent middle-class urban inspirations have buttressed these transformations at the periphery through a renewed interest in city building (Bose 2012, Fernandes 2004) and governance (Roy 2011), lending the idea of "world-class cities" a broader subscription. Some scholars have traced the ways in which the urban periphery is socially and spatially splintered while being networked through infrastructure that exclusively serves the enclaves of the rich (Graham and Marvin 2001), simultaneously producing new, complex and contested geographies of urban inclusion and exclusion (Dupont 2007).

However, the attempts by coalitions of state and market players to convert peripheries to urban uses have been met with mild to intense protests, violent resistance and localized political agitation by land losers and other stakeholders. These protests contest the regime of accumulation that takes place through forcible land grabs, dislocations and dispossessions, and violent expulsions from the use of land (Roy 2003, Levien 2013, Sassen 2014). These reactions have not only rendered large-scale urban projects partially unsuccessful, but have substantially impeded the speed with which such projects are conceived and implemented. While the problematic and partial transformation of peripheries has attracted attention from academics, there has as yet been little research that has explored the nature of these contestations and collaborations that shape such projects at the local level. There is almost no literature on the contingent nature of transformations—on the agency, actions and relative influence of entrenched social forces that constitute such dynamic localities.

This chapter traces how different actors on the ground perceive, appropriate, negotiate, shape and contest the globalizing agendas behind grand mega-urbanization projects. In doing so, the chapter argues that the local is not merely a static receptacle for global processes of economic and political restructuring but a force field that reconstitutes the parameters of "worlding" cities (Roy and Ong 2011). Analysing the process of uneven development of Hanoi's periphery, Labbe (2014) has argued that grassroots and local interest groups have penetrated the more porous levels of state-level bureaucracy to influence and modify the planning decisions behind development projects, countering the magnitude of dispossession of peasants. This chapter lays emphasis on examining forms of "homegrown neoliberalism" (Roy 2011, 263)—local coalitions, and their intersection with regional and global interests, to give concrete shape to the diverse trajectories of urban transformations that may emerge. Datta (2015) too has argued that local factors such as history, politics and the existing regulatory regimes have tempered and "provincialized" the global imaginings of futuristic smart cities such as in the making of Dholera, India. The Rajarhat case study illustrates that local non-state actors and territorial processes play an active and important role in negotiating and influencing the terms and conditions of rural to urban transformation, by entering into strategic but unstable coalitions with a range of state and other actors.

These negotiations, I argue, occur largely on an informalized and deeply politicized terrain of urbanization. One distinctive aspect of this is how the state government itself adopts and implements new governance practices in a selective and exceptional manner, mobilizes certain visions of the urban, and often circumvents its own rules and regulations in a bid to smoothen and hasten the process by which peripheries are recast into urban enclaves. This case contributes to emerging theorization about peri-urban transformations by illustrating the ways in which localized forces assert political weight to mediate and modify attempts by the market to restructure territories according to the global and national imperatives for the purpose of capitalist accumulation—a process in which the state is deeply implicated (Roy 2010).

The idea of "perforations" has been used in this chapter to understand how grand master plans are ruptured, altered, tweaked and constantly redrawn, producing a shifting geography of contested claims at the periphery.

These claims constitute complex sets of negotiations and power configurations set in motion via the process of transformation. While the media has highlighted struggles waged by the peasantry against the expropriation of their land, the reality on the ground is far more complicated. The emergence of counter struggles, divisive politics based on socio-economic status, and antagonistic class, caste and ethnic relations within land losers (Levien 2013), as well as the emergence or consolidation of local powers or economic elites trying to strengthen their economic or political clout in the area, problematizes the picture. While the state and its agencies—including certain intermediary classes or actors at the local level—may devise strategies which seek to expedite the process of transformation by buying off protests, throttling media reports or applying excessive force (Levien 2013), they may also be forced to engage in processes of negotiations over a political terrain of claims based on the recognition of use rights and exchange rights over land (Chatterjee 2004). In the process, the inherently negotiated nature of development considerably decelerates projects of mega-urbanization. This phenomenon is examined through the idea of "blockades" that have stalled the Rajarhat New Town project, raising questions of the state's and the market's ability to deliver their promises given the complex alliances and contingent coalitions that are formed between actors, institutions and agencies at the subnational and local level. There is, therefore, a need to analyse the multiple scales at which such transformations are taking place and the actions of different stakeholders, to contextualize the political nature of negotiations and power struggles over the access and use of land accompanying mega-urbanization projects.

The rest of this chapter is organized as follows. In the next section, I discuss the current political and economic landscape of urbanization, and the ways in which land has become central to the process of urban transformations and political mobilization in the state of West Bengal in Eastern India. I highlight how market reforms in the 1990s paved the way for large-scale expropriation of land, particularly on the eastern fringes of Kolkata. Next, I examine the planning of Rajarhat New Town, which reveals how the state machinery and its corollaries created exceptions and exemptions that circumvented existing town planning and land use regulations to produce "perforations" in the master plan. This is followed by an analysis of the multiple forms of land-based, anti-dispossession, people's struggles and counter-violence by the state that emerged from the contentious land acquisition and negotiated the land assembly process in Rajarhat—the "blockades". I conclude this chapter by arguing that these protests have unsettled the global imaginary of a world-class city and urban aspirations of the political elite and middle classes, by considerably slowing down the process of new town formation and introducing a degree of uncertainty and unpredictability as to when infrastructure works would be completed and plots handed over for possession. This chapter is based on ethnographic fieldwork and secondary data collection in Rajarhat between 2007 and 2009. Unstructured and semi-structured interviews with villagers, residents and officials from the Housing and Infrastructure Development Corporation (HIDCO) were used to gather data.

New towns and the political economy of urbanization, land and capital accumulation

Several new town projects of varying scales and sizes were initiated across the country after the 1990s, taking advantage of the favourable political economic climate around urbanization.[2] Most of these new township projects were initiated, planned and implemented by powerful consortiums of private developers. In addition, they were underwritten by incentives and subsidies by the respective regional state governments (Roy 2010), and facilitated through deregulations and new laws being rewritten as exceptions to existing town planning laws at the state level. Conceptualized as exclusive, gated, urban settlements promising "world-class" amenities, these new towns deliberately bypassed the existing cities and targeted a particular set of urban actors—namely the emergent upwardly mobile middle class (Bose 2012)—as future inhabitants. Following this, private developers amassed tremendous quantities of land to build more integrated townships. According to Shatkin (2011), "as large scale profit oriented urban entities, these projects represent a vision for the wholesale transformation of the urban experience through the whole commodification of the urban fabric" (Shatkin 2011, 76).

These new towns are distinctly different from the state-led planned satellite townships that were built in the 1960s or 1970s. Private players had a limited role in the actual planning and design of such townships, or in assembling land, although they were involved in influencing the decisions behind the location of new townships (Shaw 2004, Dasgupta 2007). These towns were conceived, planned and executed by state governments as strategic counter-magnets to decongest existing large cities. Urban planners had thought about the role of such towns within the larger metropolitan region and to housing different social classes. Most of these towns eventually formed their elected urban local bodies to govern and address the concerns of the residents (Shaw 2004).

The making of Rajarhat New Town has to be understood within the context of the structural transformations taking place since India's liberalization in 1991 and the concomitant political economic shifts taking place within the state of West Bengal. India's engagement with the urban, particularly building mega-urban projects and carrying out large-scale urban redevelopment projects, intensified post liberalization through a series of public–private partnerships and market-oriented reforms.[3] Land was transformed into a marketable commodity for global capital through a series of land reforms and new legislations in India.[4] Reforms outlined at the central government level included the repeal of the Urban Land Ceiling and Regulation Act of 1976, relaxation of laws for conversion of rural land to urban land use, and rationalizing stamp duties and registration processes for land sale or transfer (3iNetwork IFDC 2009). With a proliferation of actors and coalitions interested in capitalizing on land, claims to land (particularly in the periphery of urban areas) have now begun to multiply, leading to "conflicting stakes" and the emergence of a highly uneven and splintered geography of urban development (Dupont 2007).

Land and urban development are state subjects in India, and are controlled and regulated by the regional state governments. In their hands, land has become a financial instrument for leveraging infrastructure development projects, especially through the power of eminent domain. Regional state governments have been implicated in the rapid and irreversible transformation of urban peripheries by acquiring land at cheap prices from farmers and land owners and then selling these lands at very high rates to private players (Levien 2013). To build new townships or SEZs, state-level governments have used the antiquated Land Acquisition Act of 1894, which gave little or no consideration to rehabilitation and resettlement of affected persons. Moreover, the Act forced acquiescence and mobilized public sympathy by claiming that land was being taken for the "greater public good" (Searle 2010). However, in reality, these projects would expropriate land and commodify it to generate resources for further infrastructure development. These "regimes of dispossession" (Levien 2013) across various states in the country forged a powerful, national-level mobilization of opposition and protest, leading to the scrapping of the 1894 Act. It was replaced by a new law in 2013 (Right to Fair Compensation and Transparency in Land Acquisition, Rehabilitation and Resettlement Act, 2013 or LARR Act, 2013) that emphasizes higher compensation for farmers and ensures measures for rehabilitation and resettlement (Levien 2013).[5]

"Their houses on our land": the intrusion of the city into the village

The making of New Town Rajarhat is located in a particular historic moment of urban transition—Kolkata's economic decline post-Independence and its subsequent attempt at revival post 1990s as a powerful urban centre in Eastern India. Kolkata had been neglected as a city until the 1980s, since the focus of the then ruling Left Front Government was on rural land reforms and redistribution. Even when economic liberalization and reforms were being undertaken at the Centre, the subnational West Bengal government continued on its pro-poor, anti-privatization stance. Particularly with respect to land reforms, such as the repeal of the Urban Land Ceiling and Regulation Act of 1976, the state government in West Bengal refused to repeal it, fearing that an unregulated land market would alienate and exclude the urban and rural poor. In the mid to late 1990s, however, the ruling Leftist political regime recast Kolkata and its peripheries as the site for major urban transformations. The focus shifted to urban reforms and the creation of the millennial city (Roy 2003), in line with the state government's new economic policies of "privatize or perish". New actors (private developers, housing finance agencies and foreign architecture firms) entered the field of urban development and housing (Bose 2012). Though the concept of Rajarhat New Town, shaped by ideas of state-led planning, emerged in the early 1990s, its current form has been greatly influenced by these shifting relations between the state and market, as entrepreneurial forms of governance and reliance on financing infrastructure through the commodification of land began to take root in West Bengal.

As the state government began to selectively align itself with the market forces while maintaining its populist identity, it became entangled in the controversial and contradictory nature of urbanization. Brutal evictions, displacement of hawkers and removal of slum settlements from the city centre coincided with the building of mega-urban and infrastructure projects in the peripheries during this period (Roy 2003). Market-based reforms in the housing sector led to the development of the real estate sector, driven by disposable incomes and the influx of various forms of private capital (Sengupta 2013). Yet, at the same time, poorer urban and rural dwellers were marginalized and impoverished by the mega-urbanization projects that excluded, displaced and dispossessed them (Roy 2003). The state government used the logic of "territorial flexibility"—put to use by creating legal ambiguities, "unmapping" peripheries, making exceptions to planning regulations (Roy 2003), and appropriating land and using it for mega-urban development projects. Ironically, in the past, the state government had used these very same instruments to vest land in excess of ceiling limits with the urban poor, with the party functionaries at the local level controlling and mediating the use of these lands. The Left Front Government thus shifted drastically from its stated populist goals of protecting the poor and embraced the market, focusing on the radical transformation of rural land to more productive and urbanizable uses (Roy 2003, Bose 2012).

Though this form of flexible planning gave the state government a degree of control over the fate of development of the peripheral areas, it also proved to be a major bottleneck in the urbanization process. Since the state government had to exercise its powers of eminent domain and violently expel and evict people from the city's more productive spaces and appropriate commons, it led to fierce resistance. The period from 2000 to 2011 was marked by turbulent political unrest, protest from farmers and the subsequent strengthening of the opposition party, the Trinamul Congress (TMC), as it rallied forces against the forcible acquisition of land for industrial and mega-urban project development. The constant pressure of resistance and contestation, building over development projects across the state, manifested in the changing contours of the New Town project. The political mobilization of land losers against the state-sponsored regime of dispossession[6] in Singur and Nandigram was instrumental in the massive defeat of the long-established Left Front Government in 2011 (Bose 2012). After TMC came to power in 2011, it removed township development as an exempted category from the West Bengal Land Reforms Act, thus restricting private ownership of land to 24 acres, and crippling many of the large-scale urban development projects in the state which had already been affected by a sharp economic downtown in real estate across the country. Extensive lobbying in the past three to four years by the private developers and real estate lobby has now resulted in an amendment to the West Bengal Land Reforms Act by reintroducing township development in the exempted category (Datta and Basu 2014). These larger political and economic upheavals were immensely important in framing the politics of urban transformation and the rise of multiple claims to land and its use in the making of Rajarhat New Town.

Rajarhat New Town is located to the north-east of the existing city of Kolkata, outside the Kolkata Metropolitan Area (KMA) boundary. The township is an urban continuum to the electronic complex of Salt Lake city (a satellite township planned and developed by the state government in 1960s). Since its inception in 1993, a number of global and regional IT firms, real estate development companies, higher educational institutes, hotel chains and several big malls have come up in Rajarhat New Town. The bulk of housing in the township is in the form of high-rise apartments (Sengupta 2013). Very few public amenities have been built, and the dearth of facilities—particularly public transport—has deterred many people from living in New Town (Datta 2009). Current residential population estimates of the township vary from fifty thousand to a hundred thousand persons. A majority of the new inhabitants work in the IT sector or other service industries that boomed post 1990s.

The township was built over agricultural land, wetlands, water bodies and villages by displacing farmers, fishermen and villagers. However, there are still sixteen village pockets within the planned township (HIDCO 2008)—remnants of a way of life that has been rapidly effaced through peri-urbanization, and a testimony to the incompleteness and unevenness of planning. These villages have been excluded from the planning process and have also suffered physical isolation as a network of canals cut them off from the planned parts of the township. During fieldwork, one villager pointed out to the distant high-rise apartments and lamented about "their houses on our lands," encapsulating the deep-seated material and social deprivation, loss and dispossession accompanying the rural to urban transformation (Kundu 2010).

There is a history of large-scale urban projects in the eastern fringes of Kolkata. Given that Kolkata city was expanding to the east through unplanned subdivisions and encroachment of government lands,[7] the state government decided to stem the "haphazard" urbanization (Sengupta 2013) by developing two new urban settlements, Salt Lake and Kalyani in the 1970s. However, during this period, a strong regulatory regime controlled the supply, subdivision and sale or transfer of lands in these planned townships, and the entry of private players was strictly regulated by the state (Sengupta 2013). The eastern periphery of the city continued to be the preferred site for urbanization post 1990s (Dasgupta 2007). Apart from Salt Lake, the area comprised sparsely populated cultivable lands, wetlands, water bodies, villages and unauthorized settlements—basically areas that would hardly mobilize a strong protest against displacement. Thus the urban fringe presented an opportunity to rewrite the territory with a brand new story of transformation authored by the land brokering state (Roy 2003). The urbanization of the eastern periphery was in contravention to the land use development and control regulations prepared by the state itself, given the ecological sensitivity of the region (Roy 2003, Dasgupta 2007). Scholars (Bose 2012, Sengupta 2013) have thus argued that the urbanization of the eastern periphery reveals the metamorphosis of the state to an entrepreneurial one: one that capitalizes on the appropriation and sale of public lands and uses land as a financing instrument for urban infrastructure development, with real estate development at the heart of economic activity.

Goldman (2011) has studied the transformation of Bangalore's peripheries and the incorporation of dozens of villages as per the world-city master plan. He attributes the generation of exponential capital gains to the conversion of undervalued rural and urban economies to urban development projects, and argues that it is not infrastructure projects but the high value in the form of real estate development that attracts footloose investors (Goldman 2011, 246). In fact, Roy (2003) has argued that millennial Kolkata has been reshaped by speculation in real estate. This is further borne out by the "speculative" (Goldman 2011) nature of investment-driven development of New Town. Unlike earlier townships of Salt Lake and Kalyani, the West Bengal state allowed the outright sale of developed land in Rajarhat for the purpose of building urban infrastructure, to make the town planning model a "self-financing" one (Sengupta 2013). Farmers, however, were paid a meagre sum for their land, far removed from the market price and the final urban value of the developed land. Rajarhat land was acquired from farmers in the range of Rs. 7,000 to Rs. 15,000 per *katha*,[8] but it was eventually sold at an estimated price of Rs. 100,000 per *katha* (Dey Samaddar, and Sen, 2013)—at fourteen times the acquired price, indicating how the state government was able to capitalize on land sales.

What were the tactics by which the state government expelled people from their lands and livelihoods at such meagre rates of compensation without stirring up resistance? This price of land was arrived at through a set of carefully orchestrated negotiations between the state and land owners.

In fact, the state government established a base price first by directly purchasing land in Rajarhat from the farmers[9] through the formation of Land Procurement Committees (Mitra 2002). These Committees comprised top-level bureaucrats, village leaders and local political leaders from all parties. Only 10 per cent of the land mass was purchased through the Committees which involved direct negotiations. These Committees, though not given sanction by law, were immensely powerful in lending the process of land acquisition and land use conversion a degree of social legitimacy at the ground level. These paralegal committees reinterpreted the existing law on land acquisition to ensure that even tenants and farm labourers received a percentage of the sales proceeds (Chatterjee 2004, 73), thus quashing possibilities of protests and dissent.

Exceptional planning or planning through exceptions?

How was the Rajarhat New Town project planned and executed? How was the state able to negotiate with the multiple demands, while at the same time consolidating legitimacy for a "world-class" township? Gururani (2013) has used the concept of "flexible planning" to describe a range of political techniques through which Gurgaon city, on the fringes of Delhi, was built—through exemptions, exceptions and appropriations. In studying the making of Rajarhat New Town, I borrow from this understanding of urban planning as an open-ended process that is deeply politicized and informalized through a state of deregulations (Roy 2010). This approach underscores the negotiated nature of planning, spanning across state and non-state actors. Roy (2010) has argued that in the specific context of India, planning itself must be viewed as a highly informalized activity in which the state is deeply implicated. In Rajarhat, similarly, deregulation was a key mode through which land was amassed, converted to other uses and assigned arbitrarily to different claimants and buyers. In this section, I discuss how planning is practised through a series of exceptions—a) by changing the rules of business to allow a different department to take charge of the township project; b) by a flexible application of the planning principles and constant ad hoc amendments to the land use breakups; and c) by bypassing the requirements for impact assessments.

The Left Front Government conceived of Rajarhat in the early 1990s as a rational measure to check the "unplanned and haphazard urbanization" in Kolkata's peripheries (HIDCO 1999). Instead of the Urban Development Department, the West Bengal Housing Board (WBHB) was given the responsibility of planning and setting up the new town. This was facilitated through a special decree and by "changing the rules of business" (HIDCO 2012) as encapsulated in the existing Town and Country Planning Act of 1979. This shifted the activity of planning and development of the

New Town to the Housing Board, which had far less experience in planning and executing development projects of this magnitude. In 1993 a technical committee was set up to make the initial Concept Plan, based on which a task force was constituted in 1995. This task force involved reputed planning and engineering institutions to submit a project report that included a master plan for the new town. In 1996, IIT Kharagpur submitted the structure plan for the New Town. With the identification of the actual site and its notification as a planning area in 1995, direct purchase of land started. Public land acquisition started in 1996 (as per the West Bengal Town and Country Planning and Development Act of 1979). The legal planning and development authority WBHIDCO was established in 1999, much after the land acquisition and displacement had begun. In 2000, an Environmental Impact Assessment was submitted as *fait accompli*. Thus, Dey, Samaddar, and Sen (2013) have argued that the parastatal entrusted with the project applied rules and norms selectively and post facto in legitimizing the new town project in the eyes of law and society.

The *planning principles* that undergird the state-led master plan of Rajarhat New Town were significantly altered and manipulated by the prevailing social forces, emerging market dynamics and the monopolistic, rent-seeking behaviour of some state agencies. The township was designed to be a self-sufficient urban growth node (HIDCO 1999). Planned for a million people, it was to span 3075 hectares and would be divided into three Action Areas and a well-defined Central Business District (CBD) (see Figure 7.1). While the primary land use category would be residential, the plan envisioned more than adequate open spaces and greens, as well as space to set up clean industries (HIDCO 1999), and refrained from exclusive single-use zoning. The plan proposed to house non-polluting, non-hazardous, knowledge-based or financial businesses with major retail and commercial spaces, along with a large nature park and convention centre in the CBD. Each Action Area would have market places and smaller commercial areas along with residential blocks. Land was to be further plotted and sold to private developers who would build the housing, trunk infrastructure and social amenities, and then connect these developments to the main infrastructure that was to be developed by the parastatal organizations.

However, WBHIDCO deviated from the original land use breakup and categories. While initially the planning area was declared as 3075 hectares (1995), it grew in size to 3775 hectares, and then finally settled at 3550 hectares (2008). The spill-over effect was the declaration of a separate entity in 2007 called Bhangor-Rajarhat Area Development Authority (BRADA) overseeing the planned development and control of land use in the adjoining areas of Rajarhat. However, BRADA was dissolved in 2011 by the new government, led by the opposition party, Trinamool Congress (TMC). This was mainly because BRADA became an urbanization force in itself by giving free rein to private developers and promoters. These developers and promoters

Figure 7.1 Land use map of New Town Rajarhat

Source: Obtained from HIDCO 2008

dictated and inflated land prices, influenced the norms and standards for development control, negotiated with local land owners and village councils into buying land at exorbitant prices, and even strong-armed HIDCO into providing and extending infrastructure into areas which were outside its jurisdiction.

Rajarhat was initially projected as an "eco-friendly green city dotted with trees, gardens, water bodies, providing social infrastructure, better education and hygienic environment, markets, cultural and recreational centres, sports complex, technology park, shopping complex," and so on (HIDCO 1999). However, the development of the township came at the cost of the natural ecological system of wetlands (Dasgupta 2007). Though the Concept Plan and the Environmental Impact Assessment Plan had stressed that the township would provide open spaces and preserve and create water bodies in the project area, a number of water bodies were filled up to create contiguous parcels of developable land (Dey, Samaddar, and Sen 2013). While local-level developers, speculators, political henchmen and the powerful *bheri*[10] owners were party to these conversions, the state government was also deeply implicated, as it allowed the conversions to take place against its own regulations.

The goalposts for land use breakup kept shifting as market interests and local politics intervened in the planning and development of the township. In the Concept Plan, 47.6 per cent of land was designated for open spaces. Later, in the 1999 master plan, the designated area for open spaces had shrunk to 28 per cent. The area for open spaces was further reduced to 24 per cent of the total land area. While the area for open spaces was halved, the area under commercial use and under residential use was increased, these being the more lucrative sectors for private developers. A new land use category was introduced to accommodate the specific demands of IT companies and IT enabled services (Dey, Samaddar, and Sen 2013). According to Sengupta (2013), the flexibility with which the real estate sector has been approached shows a shift in the governance paradigm in West Bengal.

As a result of these constant negotiations around the size of the township, the relative proportion and location of land uses, the loosening up of regulations and the redrawing of boundaries, the urban and rural poor and the informal sector were excluded. This occurred in several ways. Initially, the planning agencies claimed that since the area was sparsely populated, there would be minimum displacement. Civil society reports have contested this claim. The area was inhabited by poor and marginalized Dalit and Muslim families who not only practised cultivation but were equally dependent on the water bodies for their livelihood (Dey, Samaddar, and Sen 2013). Thus, an estimated 130,000 people lost their land and livelihood, their impoverishment being exacerbated by the dismal rate of compensation (Sengupta 2013) offered, and the fact that the average land holding size per household was less than 3 acres or 12,140 square metres (Mitra 2002). While HIDCO instituted committees to look into a fair resettlement and rehabilitation policy, there was very little follow-up on the ground and according to several respondents the powerful leaders of the villages, strong men, or political goons and officers in the District Collectorate controlled the final list of beneficiaries, leading to a new politics of selective distribution of benefits, divide and rule, and fragmentation of communities.

The middle class, the target residents of New Town, have been variously short-changed as well. Since land values in the notified planning area began to appreciate, private developers exploited the maximum permissible developable area to build apartment blocks in the New Town. Interviews with a senior planner in IIT Kharagpur, who was involved in the preparation of the Concept Plan, revealed that many of the reserved land parcels (for parks, gardens, local markets and community halls) that were indicated in the detailed Action Area plan layouts were misappropriated by private developers. Not only did this reduce open spaces within the planned area, it also created hardships, as the lack of markets in the neighbourhoods forced residents to buy overpriced vegetables, fruits and fish from the malls, or travel to the neighbouring Salt Lake for groceries. Moreover, new residents to the township are opposed to the thriving informal shacks and roadside vendors, although they are dependent on these shops for daily needs. This frustrates the aspirations and expectations amongst the middle-class residents for an "orderly" and "planned" environment, which they find diluted and encroached by informal vendors (*The Telegraph* 2013).

The flexible planning approach has increasingly excluded the poorer groups from accessing affordable housing. According to the initial Concept Plan, people from all income groups were to be housed in the township. Public–private partnerships were formed to produce housing in the new town. Sengupta (2013), however, points out the paradox in the housing policy, suggesting that it is caught between the dual and contradictory goals of fostering capitalist interest in development while trying to secure welfare measures. This has resulted in a "highly uneven terrain for state-market interaction" (Sengupta 2013, 357) in Rajarhat New Town. Provision of land and housing for the poor is therefore negligible. According to Sengupta (2013, 365), "the state's role, when competing uses vie for allocation, favours revenue generation at the expense of opportunities for the poor," as housing for the economically weaker sections now constitutes only 15 per cent of the total housing output in New Town.

Planning in practice in New Town Rajarhat is therefore best described as negotiated development, an ongoing tradeoff captured in the words of the Chief Consulting planner of HIDCO—"canals have been shifted, roads have been twisted. In our day to day planning of New Town, we are constantly changing the alignments based on negotiations that happen between land owners, politicians and HIDCO officers. It is a challenging exercise" (personal interview 2008). The argument put forward here is that the production of Rajarhat New Town happens along a terrain of informaliza-tion of planning by the state itself—by its attempts to bypass its rules, create exceptions to its regulations and weaken existing social relations around land in the Rajarhat with the creation of extra-legal committees, and by encouraging the entry of the private sector. In the next section, I examine how these complicated manoeuvres by the state government to control and deregulate the land and urban development processes are further challenged by the politics around urbanization it encounters and contributes to.

Perforations and blockades

The market-driven logic of the current mode of urbanization in West Bengal has been enabled by the entry of new actors and the cobbling together of new constellations of state and non-state actors, bypassing of existing gov-ernment bodies as well as established networks of local leadership and gov-ernments, creation of extra-legal spaces of negotiation around land sale and purchase, and the deregulation of planning. These processes have not only upset the party's rural voters, but have also disrupted the existing informal arrangements between party functionaries, local influential political leaders, small developers, village-level leaders, small and big businessmen, bureau-crats and administrators, and even the more powerful large land holders in the vicinity of the metropolitan area where most of these new projects are being constructed. The material and socio-political manifestations of these coalitions of interests respond to the socio-spatial transformations through particular perforations and blockades. The neat, coherent and rational land use plans drawn up by the planners in the last twenty years have been repeatedly subverted and ruptured. The plans had clearly defined areas with regular boundaries for specific land uses or activities. But on the ground, the reality is rather "messy" with zigzag boundaries and incomplete networks of infrastructure. If the top-down urban imaginary for Rajarhat was the creation of a distinct space where residents can enjoy world-class ameni-ties and networked infrastructure, and where global companies can operate unhindered, the actually existing township disrupts all these imaginings.

This takes place through what I call "perforations," drawing on Benjamin's (2008) analysis of the way in which top-down master-planned spaces are challenged by the "occupancy urbanism"—a political process through which planned spaces are appropriated in parts by the poor for their own purposes. In the case of Rajarhat, perforations in the master plan are the result of the demands and assertions that various localized coalitions of actors make on the township. What appears on the ground in Rajarhat is a continuum of formal and informal settlement patterns with many numbered streets leading up to dead ends or village clusters. Road networks are convo-luted—instead of following a legible grid iron pattern—and weave through different settlements, old and new. There exists a continuum of housing from the privately developed gated communities of the rich on village lands, to squatter settlements on public lands, to high-rise enclave-style developments in planned urban layouts. There is a distinct patchiness to the spatial outcomes that reflects the multiple ways in which urbanization occurs in the fringes.

A classic illustration of perforations is the existence of village clusters in the midst of the planned township which the planners tried to "integrate" by designating them as "service villages" (Roy 2005) in the master plan. This was in keeping with the idea that the newly planned areas would be inhabited by the exclusively upwardly mobile, white-collar professionals in the service sector. These new inhabitants would be attracted by "world-class" urban infrastructure, but would be dependent upon a whole range of informal service providers—from house maids to car cleaners, security watchmen, drivers, cooks and cleaners. Planners thought that the land losers would thus seamlessly transition into the role of the informal service providers in the changed landscape, and adopt new ways of urban living.

However, this deliberate attempt by the state to erase forms of local settlements and ways of living contradicts the very production of these spaces. Urban villages serve as reminders of the violence and mediated negotiations that took place in the making of Rajarhat New Town. These villages remained because of the active intervention of a network of local politicians and influential local-level leaders, builders, businessmen and inhabitants who pressurized HIDCO through various platforms such as the Land Procurement Committee. In places, protracted dialogues with protesting villagers were an everyday activity of HIDCO officials where boundaries of land to be acquired would be deliberated upon and compensation prices renegotiated (personal interview with HIDCO official, 2009). As a result, HIDCO had to grapple with an irregular project boundary instead of a neat outline, since it acquired only farm lands and not homestead lands. This made the provision of "world-class" infrastructure and a seamless grid iron pattern of road and residential network (considered as the most rational and efficient planning designs for human settlements) rather difficult, while sharpening the social and spatial boundaries between the existing and new inhabitants.

The urgency with which the peri-urban edge is being deregulated and brought under state-market alliance has been met by a series of protests and resistances—both organized and sporadic—as well as physical and symbolic acts of violence that render the space of New Town unstable and open to multiple forms of appropriations. These are what Roy (2011) calls "blockades"—the contingent local conditions that slow down urban development and point to the inherent limits of neoliberal urbanization. Levien (2013) has discussed in detail the conditions under which strong resistance to land dispossession emerges—and more importantly, also when it does not or else is too weak, fragmented or simply broken down through

Figure 7.2 A residential cluster with a billboard advertising a world-class way of living; an erstwhile villager passes by on his bicycle

Source: Author

counter tactics—presenting a heterogeneous field of dispossession politics. In Rajarhat, land losers have used a wide range of bottom-up, informal urban strategies—such as negotiating with developers, influencing land surveyors, actively resisting land acquisition or informally occupying newly created planned spaces. These blockades have in turn shaped the perforations in the current land use patterns in New Town.

While Action Area I was acquired and developed without too many disruptions, engineers and land surveyors of HIDCO have said that this was not the case for the remaining Action Areas where land loser collectives had mobilized against the state government and halted surveys of their lands by putting up physical barriers to their villages.[11] There are several reasons why there seemed to be no overt opposition at first. Civil society organizations have claimed that the initial protests by farmers were violently put down by the Left Front Government and other political agents, larger land owners and land mafia operators, as well as local-level developers, who facilitated the conversion of land use and transfer of ownership through threats, intimidations, spoiling agriculture produce, cutting off water supply

139

to farms and filing false cases against land owners who refused to part with land (Majumdar and Nilayan 2009).

The lack of protests in the initial phase may also be attributed to the contingent coalition of local interests—local party officials across the ruling and opposition factions were involved in the expropriation of the farm lands. The Land Procurement Committees that were formed by the state government through the promulgation of a special Ordinance to negotiate the price of land with landowners in Rajarhat co-opted political leaders from across the political divide (Chatterjee 2004, Jawed 2009). These procurement committees were unique and had "extra-legal" standing, meaning they were created by the state but not required by law (in this case, the operative Land Acquisition Act of 1894). This indicates how the state government was stretching beyond its mandates and making an exception to the law in facilitating the process of expropriation of land from farmers. These committees negotiated the price of land at the ground level on a case-by-case basis, thus establishing a base price for land. The committee was deemed necessary given the complex and often undocumented tenancy structure and lack of clear titles with respect to land. The committee further declared that part of the compensation to be given to the land owners would be distributed to tenants and land labourers, who would also lose their livelihoods due to the acquisition. As per the Land Acquisition Law of 1894, no such provisions are legally available to unregistered or undocumented tenants or labourers, thus indicating the power wielded by the committee that was able to flexibly interpret the rule using several exceptional measures. However, Chatterjee (2004) has cautioned that these "extra-legal" committees were possibly shot through with existing unequal power relations in society, and therefore functioned coercively in ways to perpetuate those very inequalities—especially those based on social hierarchies created by the interplay of communal, caste and class relations, access to political connections and the material claims to land ownership and its use. Their unique position and power—having an institutional home and government mandate yet remaining outside the existing framework of the Land Acquisition Act of 1894—permitted the committees to negotiate and fix the price of land and manipulate terms and conditions through which land would be expropriated and livelihoods rehabilitated over a deeply contested political terrain. Those who served on these committees, such as local leaders and panchayat members, became even more powerful by leveraging their knowledge of land transactions and prices as well as exploiting their proximity to the West Bengal Housing and Infrastructure Development Corporation (WBHIDCO) and the builders and developers, and their knowledge of the local political dynamics and valuation of land.

According to Levien (2013), land struggles and social movements around land dispossession have to use multiple tactics—physical, political and legal. While in the initial decade of the New Town, opposition to land acquisition was fragmented, post 2000 it gathered momentum, rendering anti-land acquisition an election-worthy agenda. However, when the rising revolt against the Left Front Government's strategy of real estate led development over Singur and Nandigram, disgruntled land losers in Rajarhat began to slowly mobilize, organize themselves, refuse compensations and block planned development by cultivating on lands they had been forced to give up (Jawed 2009). The opposition party too built on these sentiments with declarations of supporting poor farmers who had been duped and dispossessed by the ruling party (*The Telegraph* 2010), consolidating its political stake in the periphery of Kolkata.[12]

A plethora of court cases against HIDCO's land acquisition process served to considerably slow down, weaken and dilute the progress of work in Rajarhat. The gains of land development were offset by the huge cost overruns due to the protracted process of mobilizing and assembling land, court cases, and the problem of laying infrastructure networks such as electric poles and water pipes with significant pockets of villages in between. Roy (2010, 87) thus asserts that while informality makes possible the

Figure 7.3 A tea stall for construction workers and villagers located on the Main Arterial Highway with the apartments in the background. The flag in the foreground claims allegiance to the Trinamul Congress party

Source: Author

territorialized flexibility of the state, it can also paralyze the developmental-ism of the state in myriad "Lilliputian negotiations" that emerge from the multiple claims around land.

Conclusions

The urbanization of Kolkata's periphery has been a messy process with several contradictory claims and intersecting scales of governance, animated by a range of aspirations for urban living. The case reveals how grassroots resistance and opposition have significantly disrupted and altered the rigid and master-planned boundaries of the Rajarhat township and weakened the political power of the ruling party. Today, Rajarhat represents various inclusions and exclusions of livelihoods, activities, settlement patterns and land use arrangements. Because New Town had to engage with existing ways of living and inhabitants with deep-rooted social, economic and political ties to the land, top-down planning, and with it the global imaginary of the urban aspirations of the elite driving the design for a "world-class city," has been resisted, challenged and modified by forces on the ground. Perforations refer to the ruptured visions and alternative socio-spatial outcomes. These perforations to the master plan are contingent upon particular conjunctures of actors who have shaped the boundary of the planning area itself. Locally embedded political pressures have been instrumental in negotiating multiple sets of contradictory claims to land and its use by selectively acknowledging some claims over others. The variety of socio-spatial outcomes on the ground thus stems from understanding the dynamics of these contestations.

In a similar vein, describing the contested urbanization of Gurgaon as an outcome of "flexible planning," Gururani has remarked that planning "is imbricated in social relations of power and actively constitutes the every-day politics of sovereignty and political authority, which are powerfully expressed through emergent and unequal urban formations" (Gururani 2013, 125). If the rural hinterland was fraught with social inequalities, stemming in particular from those who controlled access or owned land and those who did not, New Town brought with it a set of new unequal relations that manifested through uneven spatial developments. It forced the existing inhabitants, particularly those who lost access to fishing and cultivation, into casual and informal labour. It rewarded existing large land owners and local leaders with immense wealth and opportunity to exploit the emerging real estate market. It also increasingly weakened the power of village-level leaders in matters of building, construction and development of local-level infrastructure, creating unrest, conflict and political fragmentation.

Rajarhat brings to light the material manifestation of the limits and con-straints that accompany the attempts to restructure relationships around land and livelihoods in peripheries of urban areas. It highlights the chal-lenges to transforming these areas into spaces for accumulation of capital and for integration into global economic networks. It also captures the ways in which spatial planning as a key instrument is itself, in practice, constantly selecting between global imaginaries and local demands, as it embeds itself in the materiality of land, the political relations and social forces that both welcome and reject the changes being brought about.

I have argued that the state of political and economic flux in West Bengal has produced multiple perforations in the master plan for Rajarhat New Town. These perforations act as spaces for building strategic coalitions, for introducing flexibility in spatial arrangements, for bending planning laws and regulations, and for creating new spatial practices that modify the visions of "world-class" city building and produce diverse socio-spatial outcomes. These processes are predominantly centred on land and its use. The state is deeply implicated in the production of an uneven geography of valorised and devalorised spaces, as it legitimizes some kinds of development in the periphery while delegitimizing others.

Not all mega-urbanization projects proceed as planned, even when they have the full political backing of the state government. In the case of West Bengal, groups of land-dependent populations have mobilized themselves and opposed the brutal expulsion from land and the unjust and inequitable form of urban development. They have challenged the logics of urbanization imposed on them from above. However, the middle and upper-middle-class supporters of the project have decried the ways in which urban development plans have been blocked, development stalled and the vision of a world-class city incrementally tweaked to accommodate a constellation of "local" political interests. The state has become a site of bitter contestations as different agencies and state departments flex their institutional weight in the development arena, often contradicting one another with dire political consequences. It is through this treacherous urban realm that Rajarhat New Town is produced, embodying within it these multiple contradictions.

The very protracted nature of the development of the new town also challenges assumptions about the speed and ease with which mega-urbanization projects are able to transform the peripheries of urban areas. In fact, the recent consolidation of ad hoc, locally contingent and sporadic protests to more politically embedded, coherent, structured and organized protests against the state's forced acquisition of land has opened up space for new deliberations over modes of urbanization and conversion of land. These deliberations leave the future of Rajarhat unsettled, uncertain and unpredictable. The process of making new towns is thus inevitably sluggish and incremental in approach as it meanders away from the grand plans contending with daily incursions by local interest groups. This case suggests that new urban spaces are caught in the contradictions of "worlding" (Roy and Ong 2011), while constantly negotiating the complexities that stem from various coalitions of actors at multiple scales staking their claims to the emerging spaces, thus making mega-urbanization a deeply political process.

Notes

1 The Left Front Government came to power in 1977 and consolidated its rule in West Bengal for three decades through its policy of redistributive agrarian land reforms embedded in its strong Marxist ideology. In the 1990s, the state shifted its attention to urban development, wooing global and national private investors to set up new knowledge parks and manufacturing plants and build housing enclaves and infrastructure, particularly in the periphery of urban areas in tune with economic reforms at the national and state levels. Peasants began to protest the forcible acquisition of their lands and in two places, Nandigram and Singur, widespread violence erupted between the police and peasants. Although throughout the nation, farmers were protesting against the expropriation of their lands by the state for private parties, in West Bengal the protests took on an exceptionally sustained and violent form because of the smaller size of plots, lower levels of compensation and sheer number of affected farmers and farm workers. The opposition party, Trinamul Congress (TMC), vehemently opposed the land grabs and launched a political campaign against the ruling party. In 2010, the Left Front Government lost the election to TMC.

2 Amby Valley, Lavasa and Magarpatta in Maharashtra; Gurgaon near Delhi; Rajarhat and West Kolkata township in West Bengal; Mahindra World City in Chennai.

3 Several organizations, both government and private, were in agreement that India was at the point of a massive and unprecedented scale of urbanization that required building new towns, new infrastructure and massive investments. For example, the McKinsey report on India's Urban Awakening (2010) predicted that India's urban population would double by 2030. The policy of building new world-class cities was justified by the strong assertion that they could potentially generate "70% of net new jobs by 2030, produce around 70% of Indian GDP, and drive a near four-fold increase in per capita income across the nation" (MGI 2010). The McKinsey Report projected that about 200 new towns or satellite cities would have to be developed in the vicinity of existing megacities to balance economic growth and manage spatial development (MGI 2010), showing a clear alignment of interests between the state and the market in deciding the urban future.

4 In 2005, legislation was passed that allowed for 100 per cent foreign direct investment (FDI) in the development of large-scale townships of 10 hectares or more to boost the real estate and construction sectors. In addition, land use conversion from rural to urban was being made easier. The Land Ceiling Act was being repealed.

5 The LARR Act, 2014 continues to be at the centre of controversy and debate with the current Modi government at the Centre seeking an Ordinance to bypass some of the hard-won measures such as requiring the consent of farmers, conducting social and environmental impact assessments of projects, and so on. Opposition parties have launched a stiff attack against the Ordinance.

6 In Nandigram (for a chemical industry-based SEZ); in Singur (for Tata group's automobile manufacturing plant); and in Rajarhat (for the new township).

7 On the west, the river Ganges acts as a natural geographic limit to expansion.

8 In Kolkata, 1 Katha of land=720 square feet or approximately 67 square metres.

9 Although the process purportedly enabled farmers to negotiate a fair price, the resentment and violence that ensued later on, including court cases, seem to suggest that the compensation price was far from "just" and the process was possibly compromised by the threat of violence, coercion and pressure tactics. Once this asking price was arrived at and recorded in land transaction records, the government embarked on a large-scale acquisition.

10 Wetlands in which breeding of fish is permitted.
11 Interview with land surveyor HIDCO, and engineer from Development Consultants Limited.
12 Once TMC came to power, riding on the success of its anti-land acquisition agenda in 2011, the land acquisition process and institutions governing the planning and development in and around Rajarhat, such as HIDCO and BRADA, came under severe scrutiny with the new Chief Minister Mamata Banerjee, who promised to return land to aggrieved farmers—a process that is yet to see fruition.

References

3i Network Infrastructure Development Finance Company (eds.). 2009. *India Infrastructure Report. Land - A Critical Resource for Infrastructure*. New Delhi: Oxford University Press.

Benjamin, S. 2008. "Occupancy urbanism and radicalizing politics and economy beyond policy and programmes." *International Journal of Urban and Regional Research* 32(3): 719–729.

Benjamin, S., Bhuvneshwari, R., Rajan, P. and Manjunath. 2008. "'Fractured' terrain, spaces left over, or contested? A closer look at the IT-dominated territories in east and South Banglore." In *Inside the Transforming Urban Asia - Policies, Processes, and Public Action*, by D. Mahadeva (ed.), 239–285. New Delhi: Concept Books.

Bhattacharya, R. and Sanyal, K. 2011. "Bypassing the squalor: new towns, immaterial labour and exclusion in post-colonial urbanization." *Economic and Political Weekly* 46(31): 41–48.

Bose, P.S. 2012. "Bourgeoisie environmentalism, leftist development and neo-liberal urbanism in the City of Joy." In *Locating Right to City in Global South*, by T.R. Samara, Shenjing He and Guo Chen (eds.), 127–151. New York: Routledge.

Chakraborti, S. 2014. "West Bengal seeks 'Smart Green City' tag for New Town." *Times of India*, 6 July. http://timesofindia.indiatimes.com/city/kolkata/West-Bengal-seeks-smart-green-city-tag-for-New-Town/articleshow/37867094.cms.

Chatterjee, P. 2014. *Politics of the Governed: Reflections on Popular Politics in Most of the World*. New York: Columbia University Press.

Dasgupta, K. 2007. "A city divided? Planning and urban sprawl in eastern fringes of Calcutta." In *Indian Cities in Transition*, by A. Shaw (ed.), 314–340. Chennai: Orient Longman Private Limited.

Datta, A. 2015. "New urban utopias of postcolonial India: 'entrepreneurial urbanization' in Dholera Smart City, Gujarat." *Dialogues in Human Geography* 5(1): 13–22.

Datta, R. and Basu, M. 2014. "West Bengal amends Land Reforms Act to boost investments." *Hindustan Times*, 20 November. http://www.livemint.com/Politics/L0avG1zWIGh1KuzZpYMtqM/West-Bengal-amends-Land-Reforms-Act-to-boost-investments.html.

Dey, I., Samaddar, M. and Sen, S.K. 2013. *Beyond Kolkata: Rajarhat and the Dystopia of the Urban Imagination*. New Delhi: Routledge.

Dupont, V. 2007. "Conflicting stakes and governance in the peri-urban areas of large Indian metropolises: an introduction." *Cities* 24(2): 89–94.

Fernandes, L. 2004. "The politics of forgetting: class politics, state power and the restructuring of urban space in India." *Urban Studies* 41(12): 2415–2430.

Goldman, M. 2011. "Speculating on the next world city." In *Worlding Cities: Asian Experiments and the Art of Being Global*, by A. Roy and A. Ong (eds.), 246–247. London: Wiley-Blackwell Publishers.

Graham, S. and Marvin, S. 2001. *Splintering Urbanism: Networked Infrastructures, Technological Mobilities, and the Urban Condition*. London: Routledge.

Gururani, S. 2012. "Flexible planning: the making of India's 'Millennium City', Gurgaon." In *Ecologies in Urban India: Metropolitan Civility and Sensibility*, by A.M. Rademacher and K. Sivaramkrishnan (eds.), 121–123. Hong Kong: Hong Kong University Press.

HIDCO (The Housing Infrastructure Development Corporation). 1999. *The Project Report of New Town*. Kolkata: Government of West Bengal (Unpublished Report).

HIDCO (The Housing Infrastructure Development Corporation). 2008. *Strategy for Regulating Growth*. Kolkata: Government of West Bengal (Unpublished Report).

Jawed, Z. 2009. "Rajarhat land-grab model." *The Telegraph*, 7 September. http://www.telegraphindia.com/1090907/jsp/calcutta/story_11459120.jsp.

Kundu, R. 2010. *Examining the Role of Formal-Informal Nexus in the Contested Production of Periurban Kolkata*. Chicago: University of Illinois (Unpublished PhD. Thesis).

Labbe, D. 2014. *Land Politics and Livelihoods on the Margins of Hanoi 1920 to 2010*. Vancouver, Toronto: UBC Press.

Levien, M. 2013. "The politics of dispossession, theorizing India's land wars." *Politics and Society* 41(3): 351–394.

Majumdar, J. and Nilayan, D. 2009. "India's worst land grab." *Open Magazine*, 12 September. http://www.openthemagazine.com/article/nation/india-s-worst-land-grab.

Mitra, S. 2002. "Planned urbanization through public participation: case of New Town Rajarhat." *Economic and Political Weekly* 37(11): 1048–1054.

Niyogi, S. 2009. "Homes without address in swanky satellite township." *Times of India*, 3 August. http://timesofindia.indiatimes.com/city/kolkata/Homes-without-address-in-swanky-satellite-township/articleshow/4849963.cms.

Roy, A. 2003. *City Requiem, Calcutta: Gender and the Politics of Poverty (Globalization and Community Series)*. Minneapolis, MN: University of Minnesota.

Roy, A. 2010. "Why India cannot plan its cities: informality insurgence and the idiom of urbanization." *Planning Theory* 8(1): 76–87.

Roy, A. 2011. "The blockade of the world class cities: dialectical images of Indian urbanism." In *Worlding Cities: Asian Experiment and the Art of Being Global*, by A. Roy and A. Ong (eds.), 259–278. London: Wiley-Blackwell Publishers.

Roy, A. and Ong, A. (eds.). 2011. *Worlding Cities: Asian Experiments and the Art of Being Global*. London: Wiley-Blackwell Publishers.

Roy, Uttam K. 2005. "Development of new townships: a catalyst in the growth of rural fringes of Kolkata Metropolitan Area (KMA)." *Annual Conference of Housing and Urban Development Corporation Limited (HUDCO)*. Kolkata.

Sassen, S. 2014. *Expulsions: Brutality and Complexity in the Global Economy*. Cambridge, MA: Harvard University Press.

Searle, R. 2010. *Making Space for Capital: The Production of Global Landscapes in Contemporary India*. University of Pennsylvania (PhD. Dissertation).

Sengupta, U. 2013. "Inclusive development? A state-led land development model in New Town Kolkata." *Environment and Planning C: Government and Policy* 31(2): 357–376.

Shatkin, G. 2011. "Planning the privatopolis." In *Worlding Cities: Asian Experiments and the Art of Being Global*, by A. Roy and A. Ong (eds.). London: Wiley-Blackwell Publishers.

Shaw, A. 2004. *The Making of Navi Mumbai.* New Delhi: Orient Longman.

The Calcutta Gazette. 1980. West Bengal Town and Country Planning and Development Act 1979. Development and Planning (T and CP) Department. Government of West Bengal. http://www.wburbandev.gov.in/html/kolkata_gazatte.html

The Gazette of India Extraordinary. 2013. Right to Fair Compensation and Transparency in Land Acquisition, Rehabilitation and Resettlement Act 2013. Ministry of Law and Justice. Delhi. http://indiacode.nic.in/acts-in-pdf/302013.pdf

The Telegraph. 2010. "Mamata deals Rajarhat land punch." 13 November. http://www.telegraphindia.com/1101114/jsp/bengal/story_13175178.jsp.

———. 2013. "Local markets for New Town." 18 January. http://www.telegraphindia.com/1130118/jsp/saltlake/story_16453361.jsp#.VevHNhGqqkp.

The World Bank. 2013. *Urbanization Beyond Municipal Boundaries: Nurturing Metropolitan Economies and Connecting Periurban Areas in India.* Washington D.C.: The World Bank.

Part III

MEGA-URBANIZATION AND MASTERPLANNING

8

MEGA-SUBURBANIZATION IN JAKARTA MEGA-URBAN REGION

Delik Hudalah and Tommy Firman

Introduction

Over the past decades, our global economy has experienced a worldwide division of labour typified by, among other things, the shift of manufacturing operations from industrialized countries to a select number of emerging economies. It is argued that industrial globalization through spatial concentration of foreign direct investment (FDI) in manufacturing has contributed to the emergence of mega-urban regions in the Global South (Douglass 2000). A mega-urban region may consist of a mega-city, referring to a metropolitan core with more than 10 million inhabitants, and their immediate suburbs. Most mega-urban regions are located in Asia and have played a dominant role in the domestic economies where they are located (Jones 2002). They also have become a key feature of the current global system of city-regions and urban structure (Brenner and Schmid 2011).

Suburbanization is a spatial 'challenge' in mega-urban regions. The suburbs in these regions tend to extend far beyond local administrative boundaries of metropolitan core towards formerly rural areas. They often represent the most dynamic and fast-growing areas of the regions due to rapid land-use conversion into urban functions, large-scale land development, and manufacturing deconcentration (Douglass 2000, Hudalah et al. 2007).

Industrial suburbanization in a mega-urban region is generally caused by a huge demand for cheap vacant land for the expansion of industrial parks and their supporting infrastructure (Feng et al. 2008). Manufacturing industry relocate to suburbs to lower the costs of production, while the global market coordinating function of the industry is maintained in the core city (Shatkin 2008). Industrial revenue sources will depend on the global economy and cooperation with other multinational companies.

Jakarta Metropolitan Area (JMA) has become one of the largest mega-urban regions in South East Asia (Murakami et al. 2003). Since the 1990s, the population of Jakarta's suburbs has surpassed that of its metropolitan core (Hudalah and Firman 2012). In relation to this population dispersal, other important elements of Jakarta's suburbanization have entailed massive urban land-use conversion and large-scale housing development (Firman 2004). The land subdivision projects have mainly taken the form of dormitory and industrial towns. In the peak period of the 1990s property boom, more than 23 new towns ranging from 500 to 6,000 hectares were developed around Jakarta alone (Winarso and Firman 2002). Currently, there are more than 35 industrial parks in Greater Jakarta with a total area of over 18,000 hectares (Hudalah et al. 2013).

Suburbanization of economic activity in Greater Jakarta was triggered in the 1980s by, among other things, the national government's deregulation and debureaucratization measures. These market-oriented policies aimed to attract domestic and foreign private investors in finance and manufacturing industries to accelerate economic growth (Firman 2000). The policies not only boosted FDI in manufacturing, but also real estate development, leading to uncontrolled growth of large-scale projects by private land developers in the suburbs. In addition to the influence of this government-supported global capitalism, other institutional factors contributing to Greater Jakarta's suburbanization include the rise of a middle-class society, clientelist governance practices, and the weakened presence of government in the suburbs (Hudalah et al. 2007, Hudalah and Woltjer 2007).

While the physical, demographic, and economic dynamics of the fringes of Asian mega-urban regions are well documented, little attention has been paid to the role of politics in reshaping this suburbanization trend. As a grey area between a city and its rural hinterland, suburbs in mega-urban region have been associated with blurring local authorities, thus tending to be invisible from formal political frameworks and intervention (McGee 1991). There is evidence of the private actors in the suburb beginning to build relationships with government and influence the direction of government policies and plans (Shatkin 2008). This chapter will investigate the role of actors and politics in managing and directing mega-suburbanization around Jakarta. It identifies key political actors, which include public and private sectors, and their roles in influencing the decision-making processes over the spatial development of the new industrial cities in suburban Jakarta.

In this chapter, we will discuss emerging urban politics theories and how these can be used to identify suburban transformation in the mega-urban region of Jakarta. The chapter will proceed with an overview of FDI in manufacturing and its spatial implication for JMA. The case study will explore key political actors. For this purpose, we have interviewed about 20 relevant actors and conducted content analysis on archival data such as policy documents and reports, minutes of meeting, and newspaper articles. It will be argued that the influence of the capital on mega-suburban spatial changes is not autonomous but strategically intermediated by the dynamics of local, regional, and national politics. As a result, the spatial transformation of suburban JMA in the past three decades has followed three main phases: industrial-park corridor, agglomeration of corporate cities, and new regional growth centre.

From urban to suburban politics

Douglass (2000) argues that the suburbs of mega-urban region have the potential to become the sites of major contestation between global economies and local politics. In addition to physical and economic issues, political issues such as redistribution, inclusiveness, and segregation are increasingly becoming an important element of urban and suburban governance. There are two influential theories of urban politics explaining the relationships between the political and economic actors in suburban development. These theories are the growth machine theory and the urban regime theory (Phelps and Wood 2011).

Growth machine theory was first coined by Molotch (1976) to explain the political dimension of city building in the United States. According to this theory, urban development is driven by the interest of the local private elites who compete with each other mainly to obtain maximum profit. The elites are supported by government intervention in key policy themes such as labour costs, tax rates, and applicable law (Molotch 1976). Following the decision of an industrial company to set up a branch factory in a particular location, the land-use patterns of the surrounding area are formed. In addition, the spatial changes that occur in a locality will also greatly affect the rational consideration of owners of industrial enterprises to relocate their factories. Industrial relocation is then followed by an increase in industrial employment and the trade and services sectors, intensive land development, population density, and an increase in financial activities.

While growth machine theory emphasizes local elites, urban regime theory, as introduced by Stone (1993), pays primary attention to the effectiveness of formal and informal coalition between the political and economic actors in city building and management. The urban policy is no longer merely driven by the main actor's direct interests but the need to sustain the wider economic growth of a city. We do not see who has the full power over the urban life but how power is represented to achieve certain goals by considering the relationships that occur in the social life in urban areas. In this case a kind of vision is built to blend a variety of interests of all parties. Urban issues are settled through the participation of a selective distribution contract governed by the division of labour, the provision of facilities for the community, and others. Stone (1993) suggested that the effectiveness of local governments depends on their cooperation with non-governmental actors and a combination of government and non-government resources.

Phelps and Wood (2011) proposed that the two influential theories should not be contrasted but rather linked with each other to provide a framework guiding various stages of suburban land development. First, the growth machine is generally applied to relatively new suburbs where the local government is still relatively immature. It is characterized by extensive land-use conversion from rural to urban functions for the first time. If industrial activities largely characterize the suburban transformation, the political machine can continue to grow or be updated. However, if a suburb is dominated by residential areas, it is likely that the growth machine will gradually be replaced by an urban regime (Phelps 2012).

New suburbs can later transform into mature suburbs, post-suburbs, or cities characterized by a diversifying and intensifying pattern of land development (Phelps and Wood 2011). During this transformation, government intervention begins to offset the dominance of the private actors, which signifies the transition from growth machine to regime political arrangement. The objectives are no longer urban land transformation but how to establish a mechanism that retains the advantages of living in a suburb for the local population while meeting the global demands of reshaping the suburb. The local political economic interest is no longer focused on the exchange value but on the long-term use of the land (Phelps and Wood 2011).

Industrial suburbanization in Jakarta Metropolitan Area

Covering a total area of 5897 km², JMA is the largest mega-urban region in Indonesia (Hudalah and Firman 2012). It consists of *Daerah Khusus Ibukota* (DKI) Jakarta or Jakarta Capital Special Province surrounded by its suburbs known as Bodetabek, which is an acronym for Bogor-Depok-Tangerang-Bekasi. Bodetabek comprises three autonomous kabupatens (districts or rural governments): Kabupaten Bogor, Kabupaten Tangerang, and Kabupaten Bekasi. In addition, it also includes five kotas (municipalities or urban governments): Kota Bogor, Kota Depok, Kota Tangerang, Kota Tangerang Selatan, and Kota Bekasi.

Hudalah et al. (2013) argued that industrial development has played an important role in JMA suburban transformation. Government policies supporting industrial investment date back to the beginning of Suharto's New Order Regime in the late 1960s. With its market-led development policy, the New Order administration positioned industrialization as a key policy platform to boost economic growth. Industrial development was first endorsed in industrial zones (*zona industri*), referring to unplanned concentration of industrial activities. Traditionally, local and labour-intensive industries tended to locate in such unplanned industrial areas.

In comparison, foreign and hi-tech industrial investors preferred to operate in industrial parks (*kawasan industri*) (Hudalah et al. 2013). Kwanda (2000) found that in the early stages of their development in the 1970s, industrial parks were mainly built as a reaction to the increasing environmental impact and infrastructure inadequacy of the industrial zones. Industrial parks were only allowed to be built by state-owned enterprises. The state-owned industrial parks were located in the northern/coastal zone of the inner-city of Jakarta. Among others, the Jakarta Industrial Park Pulogadung (594 ha) and Kawasan Berikat Nusantara (595 ha), built in 1970 and 1986 respectively, remain in operation.

In the 1980s, the governments issued a number of deregulation packages to create a more conducive investment climate. One of them was focused on simplifying the permits procedures. For example, Bekasi District Government committed to speed up the permit procedure for building new factories to three days. This market-oriented development policy resulted in mounting proposals from domestic as well as foreign private investors who wanted to relocate their business and factories to Indonesia. The state-owned industrial parks were too limited to be able to meet this rapidly growing demand. Therefore, the government started to invite the private sector to participate in the industrial-park development. The government believed that private participation could help speed up the industrial-park development.

Many private developers had long been interested in industrial-park business but there were no rules that governed their initiatives. As the first effort, Presidential Decree 53 (1989) was enacted to provide the legal basis for private participation in large-scale industrial-park development. Another presidential decree was later issued in 1996 to set a detailed guideline for industrial-park development. Presidential Decree 41 (1996) defined industrial parks as centres for industrial activities with the provision of infrastructure and supporting facilities developed and operated by licensed industrial-park companies. The presidential decree highlighted that industrial-park development was aimed at accelerating industrialization of a region, facilitating industrial activities, directing industrial location, and strengthening environment-friendly industrialization.

Following the enactment of these presidential decrees, industrial-park development shifted from the city towards its suburbs. Since the end of the 1980s, available industrial land in Jakarta has declined. No more industrial parks have been developed in the city. Conversely, since the middle of the 1990s, the suburbs have taken over the role of the city in leading the supply of formal industrial land (Hudalah et al. 2013).

In addition to the central government's regulatory reforms, industrial-park development in Jakarta's suburbs could not be separated from the opening of two intercity highways built along the northern coastal zone of Java, which are Jakarta-Cikampek Toll Road and Jakarta-Merak Toll Road. Most of the industrial parks or planned industrial areas built since 1989 are located along these coastal highways. With the opening of these highways, the industrial parks have in the past two decades expanded beyond the boundaries of Jabodetabek. Covering a total area of 9,016.43 square km, this extended mega-urban region has recently been called 'Greater Jakarta'. There are now more than 35 suburban industrial parks in Greater Jakarta with a total area of over 18,000 hectares (Hudalah et al. 2013).

Industrial suburbanization has in turn affected the economic structure of JMA. In the past two decades, tertiary sectors (finance, service, and commercial sectors) were still consistently concentrated in Jakarta. Nevertheless, a remarkable shift could be seen in the manufacturing sector in which the suburbs' financial contribution increased substantially from only 24.6 per cent in 1985–90 to 59.8 per cent in 2000–5 (Hudalah and Firman 2012). The leading role of the suburbs in the regional economic restructuring is more apparent in the absorption of FDI. Hudalah and Firman (2012) have analysed how the suburbs have attracted most of the foreign business units in the manufacturing, real estate, and infrastructure sectors. From 1998 to 2009, the financial contribution of the suburbs in these sectors stabilized at 84–7 per cent.

Hudalah and Firman (2012) have shown that Kabupaten Bekasi, a suburban district on the eastern outskirts of Jakarta, has stabilized its role as the main industrial centre of JMA while taking the lead in the suburban economic restructuring towards service sectors. About 50 per cent of the FDI in the secondary sectors in JMA was captured by Kabupaten Bekasi. Industrial activities, particularly manufacturing, have played a key role in the development of this sub-region. In 2010, manufacturing contributed 79.73 per cent of the gross domestic product (GDP) of Kabupaten Bekasi.

The case of Cikarang, Bekasi

Industrial parks in JMA are mostly concentrated in the suburban district of Bekasi. The central government has long designated the suburban district as a buffer zone between Jakarta and its rural hinterland (Pemerintah Republik Indonesia 1976). It was mainly allocated for industrial and residential development while maintaining its function as one of the nation's rice cultivation centres.

Industrial parks in Kabupaten Bekasi are mostly concentrated in Cikarang, the central part of the suburban district. There are now seven industrial-park companies in Cikarang covering a total planning area of 14,620 ha (Table 8.1). Several of the industrial-park developers in Cikarang were established through a joint venture with foreign investors. For example, the MM2100 was established in cooperation with Marubeni Group, a Japanese investor. The Hyundai Industrial Estate is a joint venture between Lippo Cikarang with Hyundai Corporation, a Korean company. Currently 2,288 industrial companies are located in these industrial parks (and towns). These companies originate from various countries. In Jababeka alone, for example, there are more than 1,500 national and multinational companies from more than 30 countries, including the United States, the UK, France, Germany, the Netherlands, Australia, Japan, South Korea, China, and Taiwan (PT Jababeka 2010). Most of the companies are typified by hi-tech industries

Table 8.1 Industrial parks in Cikarang, Bekasi

Industrial estate/town	Industrial area (ha)	Planning area (ha)	Industrial tenants	Workers
MM2100 and BFIE	1,250	2,500	440	52,814
EJIP	220	320	87	23,142
BIIE (Hyundai)	140	200	105	27,800
Jababeka	1,570	5,600	1,235	328,510
Lippo (Delta Silicon)	686	3,000	378	100,548
Deltamas (Greenland)	*)	3,000	43	*)
Total	3,866	14,620	2,288	532,814

Source: Hudalah and Firman (2012)
*) no data available

such as automotive, electronics, chemicals, machines, metals, and plastics. These companies have attracted more than half a million workers. In addition to Indonesian workers, the data from Regional Investment Indonesia show that, in 2005, there were 3,004 expatriates living in Kabupaten Bekasi and 6,290 others living in Kota Bekasi.

Until 2009, Cikarang had a total industrial land area of 6,214.4 hectares, making it the largest planned industrial centre in Southeast Asia. Hudalah and Firman (2012) have reported that the seven industrial parks (and towns) in Cikarang have a potential export value of up to $US 15.1–30.56 billion, or about 46 per cent of the national non-oil and gas export of $US 66.428 billion (2005). This contribution is much larger than that of its runner-up, Batam industrial parks, with $US 4.6 billion. The government has also successfully extracted taxes totalling 3.4–6 trillion rupiahs from the industrial activities in Cikarang.

The changing politics of a suburb

The rapid physical transformation in Cikarang in the past three decades could be explained by the dynamics of participating actors, their interests, and the resulting governance styles. Based on these indicators, we analytically divide suburban development in Cikarang into three main phases: industrial park, corporate town, and special economic zone (Table 8.2). We will elaborate on each of these phases in the following sub-sections.

Industrial relocation in Cikarang, Bekasi

One of the government's ambitions during Soeharto's market-oriented administration was to spread the economic development by, among other things, encouraging industrialization outside Jakarta. Therefore, following the operation of this 83-km toll road in 1988, the Minister of Industry

157

Table 8.2 The dynamics of actors, politics, and governance in Cikarang

	Industrial park (1980s–90s)	*New town (1990s–2000s)*	*Special economic zone (2000s onward)*
Wider contexts	Market-oriented policy	Economic boom; rise of middle-class	Decentralization; economic crises
Governance style	Clientelism	Corporatism	Network; partnership
Key interest	Growth	Amenity and security	Sustainability and competitiveness
State intervention	Inter-regional infrastructure development; industrial land development policy	Large-scale land development policy	Special economic zone policy (KEK)
Private developer	Land acquisition; industrial land development	Urban planning; residential land development	Regional planning; regional infrastructure development
Non-governmental institution	Indonesian Industrial Estate Association (HKI)	Township management	International Zone (ZONI) Cikarang

indicated Bekasi District, especially Cikarang, as a new area for industrial expansion (Figure 8.1).

In addition to its direct access to the country's second intercity toll road, there were several other reasons for selecting Bekasi District as the new prime location for industrial-park development. First, Bekasi District is only about 40 km from Jakarta, the country's business and financial capital, where most manufacturing headquarters are located. It is also relatively close to Tanjung Priok International Seaport, the country's largest manufacturing export–import gate. Good access to water resources is also important for manufacturing activities, and this suburban district is passed by water directly channelled from Jatiluhur, the biggest reservoir in West Java. Finally, a large part of Cikarang formerly consisted of arid land, whose soils were only used for clay mining for brick production. It was considered much cheaper and justified from a planning perspective to build industrial parks on formerly unfertile soils.

Labour concentration was another consideration for transforming Bekasi District into a new industrial centre. In fact, in 1985 the Ministry of Manpower had anticipated preparing skilled labourers to support the operation of manufacturing industries in Bekasi with the establishment of a training centre.

Most industrial parks in Cikarang were developed since the late 1980s, especially following the enactment of Presidential Decree 53 (1989) explained

Figure 8.1 Map of Cikarang
Source: own analysis (various sources)

in the previous section. Major land developers first approached unprepared district officials to apply for principal permits, indicating proposed locations for their industrial parks. Once the permits were given, the developers started to acquire land. In the process of land acquisition, a developer was obliged to consult with the land owners, farmers, and local residents about a compensation payment and, if necessary, land substitute in other similar locations. As another step, the developer was required to prepare a master plan, which needed to be approved by the district government. A master plan should include detailed land-use allocation and supporting infrastructure and facilities in the industrial park. With such a private-driven mechanism, the developers had a strong influence in directing the spatial utilization in the industrial locations.

The capacity of Indonesian big developers to direct urban policy has long been recognized (Winarso and Firman 2002). In Cikarang, this active role of the developers can be seen in the design of local land-use planning regulatory frameworks:

> We must say that the [urban planning] policy is de jure made by the district government. However, everywhere what is implemented de facto is our [industrial-park] planning regulation. We force the local government to adopt our regulation.
>
> (Interview with industrial-park manager, 6 March 2012)

Building clientelist relationship with the governments, the developers took part in the policy realm so that the resulting policies could accommodate their interest in boosting regional economic growth:

> We in such a way pushed – so, the point was we pushed – the regulation to accommodate the (economic) growth, either for the national development or for the sake of company's profit. . . . It would be difficult for us, for example, to invite investors to build factories but there was no road. They surely wouldn't come. We, as developers, acted as a 'shadow' government.
>
> (Interview with industrial-park manager, 6 March 2012)

Corporate new towns

Among the seven large-scale development projects established in Cikarang, four of them, which are the MM2100, Bekasi Fajar Industrial Estate (BFIE), East Jakarta Industrial Park (EJIP), and Bekasi International Industrial Estate (BIIE), have focused their business operation purely on industrial development and management (Table 8.3). Meanwhile, three other private developers, which are Jababeka, Hyundai (Lippo Cikarang) and Deltamas, gradually envision to build "corporate towns" rather than "industrial parks" thus developing not only industrial sites but also residential areas and their supporting facilities (Hudalah and Firman, 2012).

Initially, as Hudalah (forthcoming) has indicated, the business orientation of Lippo Cikarang and Jababeka was to build industrial parks. However, since the construction of industrial parks, the population of Cikarang has grown rapidly from 175,000 (1986) to 733,000 (2010). This growing population has created massive demand for the development of residential areas and supporting facilities. Therefore, the developers later realized the industrial parks could not stand alone. The development of industrial facilities in both locations has gradually been combined with the development of middle- and upper-class houses. To ensure a secure and convenient living environment, the developers built residential clusters with a single entry point (Figure 8.2).

Table 8.3 Visions and missions of industrial parks in Cikarang

Industrial park	Vision	Mission
Jababeka	To become the most respected and environmentally friendly township developer	Reliable, professional, and always striving to exceed expectations
Lippo Cikarang	Place to work, live, and play	To become a leading industrial, commercial, and residential-based urban developer in Indonesia, through urban infrastructure, public facility, and managerial investment
MM2100	To be Indonesia's leading industrial provider of superior business environment and services for manufacturing companies and related operations	
East Jakarta Industrial Park (EJIP)	Global corporate making great contributions to society	To become a leading industrial park in Asia
Pembangunan Delta Mas	Better living standards	To create a pleasant, harmonious, and sustainable living environment for its residents
Hyundai Inti Development	To create a high-technology and comprehensive industrial park to support Korean companies	

Source: Hudalah (forthcoming)

The resulting built private environment tends to transform into a property machine that divides and rules the growing population (Merrifield 2013). It has been argued that such exclusive housing construction has facilitated the formation of gated communities, which encourage social segregation at the community and local levels and economic disparity at the regional level (Firman 2004, Hudalah and Firman 2012, Leisch 2002).

As a further development, the residential areas in Lippo Cikarang and Jababeka are also supported with premium facilities such as a private university and schools, a world-class hospital, star hotels, shopping malls, recreational parks, and entertainment centres. In Delta Mas, the development of residential clusters and supporting facilities has from the beginning accommodated the new site for the district capital.

Figure 8.2 Gated community in Lippo Cikarang

Source: author's collection

It is evident that each industrial-park developer has built an excellent infrastructure network within their own project location (Hudalah forthcoming). However, every project tends to stand on its own. It is not adequately linked to its neighbours. The industrial parks are poorly interconnected with each other. This poor physical connection has made the internal structure of Cikarang become fragmented. The resulting urban transport system is also chaotic since all regional as well as local traffic is highly dependent on the intercity toll road.

Each corporate town manager has focused their attention on improving access to the Jakarta–Cikampek Toll Road. Currently, the corporate towns have a direct link with three toll road exits. The intercity toll road was developed and is run by PT Jasa Marga, the state-owned highway company. However, most of the toll road exits in Cikarang were proposed and financed by private developers. The corporate town developers want to have a better link with the intercity toll road in the hope of increasing the accessibility of their project locations and, thus, their respective property values.

As can be found in other suburban districts around Jakarta, the local government has a weak capacity to provide adequate urban infrastructure (Hudalah et al. 2007). As a result, most urban infrastructure in Cikarang, including road and drainage networks, was built independently by the industrial-park developers. Such an uncoordinated private initiative has caused conflicts:

We have experienced that when we cooperated [with the district government], the government mostly could not accomplish their parts. There have been several disputes because of this. Clearly the bottleneck was in their parts. The upstream was in Lippo's, passing through Mulia's . . . [but] the bridge was too narrow. As a result, [Mulia] started to flood. Mulia blamed us but there was nothing we could do because the problem was on the bridge owned by the government.

(Interview with an industrial-park developer,
3 April 2013)

Towards a networked governance?

While the city continues to grow, the local traffic in Cikarang has largely depended on the intercity toll road. As a result, PT Jasa Marga, the state-owned highway corporation, has recorded that during the period of 2008–12, the number of vehicles passing through the toll road exits grew significantly from 337,000 to 532,000. The concentration of all traffic on the highway has caused acute congestion. The industrial-park managers in Cikarang worry about the impact of this traffic congestion on the decrease of land value in the industrial parks and, thus, the competitiveness of Cikarang as Indonesia's prime location for FDI in manufacturing.

Maintaining the competitiveness of Cikarang on the global stage of attracting investment has become the common interest of the seven industrial-park managers. For this reason, they agreed to establish an industrial-park association called *Zona Internasional* (ZONI). ZONI can be seen as a 'worlding practice' (Roy and Ong 2011), aimed at pursuing a better world reputation of Cikarang in the face of intercity and regional competition playing out on the global stage. ZONI was first proposed by Jababeka, the biggest industrial-park manager in Cikarang, but then approved by the others. This public–private partnership platform was signed and declared in 2006 by involving the central government (the ministry of public works), West Java Provincial Government, Bekasi District Government, Jasa Marga, and the seven industrial parks.

ZONI envisioned preparing Cikarang as an excellent and competitive industrial district. This collaborative platform was, among other things, aimed at promoting development acceleration and investment growth in Cikarang. To accelerate the economic development, ZONI carried out investment promotion joint programmes abroad.

To improve the investment attractiveness, the participating industrial-park managers conducted a feasibility study to propose Cikarang as a special economic zone (SEZ). To some extent, this proposal attempts to replicate the success of *kaifaqu*, the Chinese development zone (Hsing 2010). As the largest industrial-park manager in Cikarang, Jababeka actively played a role in the SEZ proposal. The SEZ would focus on debureaucratization and investment service integration to improve the business climate in Cikarang.

As another attempt to accelerate the economic development, the industrial-park managers have agreed to coordinate and collaborate in urban and regional infrastructure development. Since 2006, they have planned a number of road infrastructure development and improvement programmes in Bekasi, which include toll roads, connecting roads, bridges, flyovers, and underpasses (Table 8.4). Until 2009, it could be said that PT Jasa Marga, the state-owned highway corporation, the central government, and the province have fulfilled most of their responsibility in the joint road infrastructure programmes. Meanwhile, the district government could not complete half of their programmes in time. This was due to conflicts of interest in the local politics and limited financial resources, which has become a common problem faced by rural government in decentralized Indonesia (Firman 2013).

Table 8.4 also shows that the industrial-park managers could not carry out most of their agreed responsibility in time, especially to build inter-industrial-park connecting roads. Most industrial-park managers postponed the road construction because they experienced difficulties in acquiring the land.

The failure of the industrial-park managers to meet the agreed schedule has created conflicts between them as well as with the governments. As an extreme case, the local government was forced to send an attorney to settle disputes with an industrial-park manager. As a solution, the manager requested the government to allow them to build a new toll road exit in exchange for road construction.

Another conflict could be found during the road and bridge construction that would connect EJIP and MM2100. The central government had earlier finished building the bridges crossing the river that separates the two industrial parks. As part of the agreement, each industrial-park manager was obliged to construct the connecting road to their respective project site. However, MM2100 refused to complete their part. Having a more strategic

Table 8.4 Road infrastructure programmes under ZONI (2006–9)

Responsible actor(s)	Finished	In progress	No progress	Total
Central government	5		2	7
PT Jasa Marga	5			5
West Java Provincial Government	3			3
Bekasi District Government	1	1		2
Industrial-park managers		2		2
Public–private cooperation	1	2	1	4
Total	15	4	4	23

Source: own analysis (various sources)

location than EJIP, they worried about an increase of traffic and road maintenance costs in their project location following the road construction.

The lack of trust and commitment among the industrial-park managers in Cikarang originated from the fact that historically, they were closed-business entities who had their own corporate visions and strategies and tended to compete with each other. As a network organization, ZONI or the association of the industrial managers in Cikarang aimed to build mutual understanding among these competing industrial managers and enhance their communication and cooperation with the governments. As such, ZONI functions to improve the bargaining position of the industrial parks in local and regional policy-making.

It is suggested that a complex urban space in the global era results from network-based organizations such as ZONI (Lefebvre 2003, Merrifield 2013). Hudalah (forthcoming) concluded that building of trust among the competing actors has become a great challenge as well as the key success element to realize the common vision of this network. As an illustration, a developer indicated that building trust is a long-term effort requiring regular practices:

> Previously we maintained distance from each other but then [communication] was initiated. . . . Actually all of us [the seven industrial-park managers] realized that [cooperation] is important but, since we created distance, we sometimes misunderstood . . . think negatively [about each other]. Nevertheless, it was not a big problem. We just needed evidence. And time would answer that if we did it together, everything would be easier.
>
> (Interview with a general manager of Jababeka,
> 20 March 2013)

Conclusions

As one of the largest mega-urban regions in Asia, JMA has in the past three decades physically expanded from the city to its peripheral areas. Following Soeharto's market-oriented development policies of the 1980s, new residential and industrial cities, which were mostly initiated by private development corporations, have featured suburbanization in this metropolitan region. In the last decade, there has been a substantial shift of functions and activities in the new suburban cities that may lead to the emergence of new industrial-based metropolitan subcentres.

The flows of FDI at the global scale, especially in the manufacturing sector, have become a major driver of spatial transformation in JMA. In response to stricter environmental regulations, industrial investments in suburban Jakarta have increasingly been shifting from unplanned industrial zones to planned industrial cities, which are effectively managed by private developers. The role of private actors and global capital have gained importance in reshaping the metropolitan spatial structure by influencing the local, regional, and national political landscapes.

In the past decade, Cikarang has transformed into an important suburban centre in Greater Jakarta. Agglomeration of industrial parks has contributed to the high GDP of Bekasi District. A large number of population and jobs have relocated from Jakarta to Cikarang. In the initial phase, suburban development in Cikarang could not be separated from the central government policy to relocate manufacturing industries in Bekasi District as a strategy to accelerate the nation's economic growth. During the land development phase, the industrial-park managers built a close tie with the local (district) government, especially in road infrastructure development. Accommodating population growth and urban activities, the industrial-park managers also constructed corporate cities by building gated residential areas and exclusive urban facilities as their new business opportunities.

According to Ekers et al. (2012), there are three main governing mechanisms that can best explain suburbanization. These are individuals, government, and the market. It seems that the spatial structure and pattern of Cikarang have largely been driven by the market. This primarily involves large-scale planning and resource management by private developers and financial sectors that control a large part of the suburbs with a limited role in government. The private developers were able to make and implement their own masterplans in their project locations. They could also initiate the construction of new toll road exits and major urban facilities and amenities. Each industrial manager built the infrastructure independently without prior coordination with their neighbouring project managers. As a result, the project locations tend to be disconnected from each other, creating fragmented spatial structures at the local and regional levels.

The North American growth machine concept is concerned with the interrelationship between the private sector and local government. Meanwhile, it can be concluded that the public–private partnership in Cikarang has been far more complex. The private developers, with the support of foreign and domestic investors, have acted as a 'shadow government'. They have been able to strongly influence the central government's decision-making since the drafting of the first presidential decrees on industrial development. Furthermore, the developers are able to strategically influence the local government through planning policy formulation and urban infrastructure development.

References

Brenner, N. and Schmid, C. 2011. "Planetary urbanisation," in Gandy, M. (ed.) *Urban Constellations*. Berlin: Jovis Verlag, 10–13.

Douglass, M. 2000. "Mega-urban regions and world city formation: globalisation, the economic crisis and urban policy issues in Pacific Asia." *Urban Studies* 37: 2315–2335.

Ekers, M., Hamel, P. and Keil, R. 2012. "Governing suburbia: modalities and mechanisms of suburban governance." *Regional Studies* 46: 405–422.

Feng, J., Zhou, Y. and Wu, F. 2008. "New trends of suburbanization in Beijing since 1990: from government-led to market-oriented." *Regional Studies* 42: 83–99.

Firman, T. 2000. "Rural to urban land conversion in Indonesia during boom and bust periods." *Land Use Policy* 17: 13–20.

Firman, T. 2004. "New town development in Jakarta Metropolitan Region: a perspective of spatial segregation." *Habitat International* 28: 349–368.

Firman, T. 2013. "Territorial splits (Pemekaran Daerah) in decentralising Indonesia, 2000–2012: local development drivers or hindrance?" *Space and Polity* 17: 180–196.

Hsing, Y.-T. 2010. *The Great Urban Transformation: Politics of Land and Property in China.* Oxford: Oxford University Press.

Hudalah, D. Forthcoming. "Governing industrial estates on Jakarta's peri-urban fringe: From shadow government to network governance". *Singapore Journal of Tropical Geography.*

Hudalah, D. and Firman, T. 2012. "Beyond property: industrial estates and post-suburban transformation in Jakarta Metropolitan Region." *Cities* 29: 40–48.

Hudalah, D., Viantari, D., Firman, T. and Woltjer, J. 2013. "Industrial land development and manufacturing deconcentration in Greater Jakarta." *Urban Geography*: 1–22.

Hudalah, D., Winarso, H. and Woltjer, J. 2007. "Peri-urbanisation in East Asia: a new challenge for planning?" *International Development Planning Review* 29: 503–519.

Hudalah, D. and Woltjer, J. 2007. "Spatial planning system in transitional Indonesia." *International Planning Studies* 12: 291–303.

Jones, G. W. 2002. "Southeast Asian urbanization and the growth of mega-urban regions." *Journal of Population Research* 19: 119–136.

Kwanda, T. 2000. "Pengembangan kawasan industri di Indonesia." *Dimensi Teknik Arsitektur* 28: 54–61.

Lefebvre, H. 2003. *The Urban Revolution.* Minneapolis, MN: University Of Minnesota Press.

Leisch, H. 2002. "Gated communities In Indonesia." *Cities* 19: 341–350.

McGee, T. G. 1991. "The emergence of Desakota regions in Asia: expanding a hypothesis." *The Extended Metropolis: Settlement Transition In Asia*: 3–25.

Merrifield, A. 2013. "The urban question under planetary urbanization." *International Journal of Urban and Regional Research* 37: 909–922.

Molotch, H. 1976. "The city as a growth machine: toward a political economy of place." *American Journal of Sociology*: 309–332.

Murakami, A., Zain, A. M., Takeuchi, K., Tsunekawa, A. and Yokota, S. 2003. "Trends in urbanization and patterns of land use in the Asian mega cities Jakarta, Bangkok, and Metro Manila." *Landscape and Urban Planning* 70: 251–259.

Pemerintah Republik Indonesia 1976. *Instruksi Presiden Republik Indonesia Nomor 13 Tahun 1976 Tentang Pengembangan Wilayah Jakarta-Bogor-Tangerang-Bekasi [The Presidential Instruction Number 13 Year 1976 on the Regional Development of Jakarta-Bogor-Tangerang-Bekasi].* Jakarta: Pemerintah Republik Indonesia.

Pemerintah Republik Indonesia 1989. Keputusan Presiden Nomor 53 Tahun 1989 Tentang Kawasan Industri [Presidential Decree No. 53/1989 on Industrial Parks]. Jakarta: Pemerintah Republik Indonesia.

167

Phelps, N. A. 2012. "The growth machine stops? Urban politics and the making and remaking of an edge city." *Urban Affairs Review* 48: 670–700.

Phelps, N. A. and Wood, A. M. 2011. "The new post-suburban politics?" *Urban Studies* 48: 2591–2610.

Presiden Republik Indonesia 1996. Keputusan Presiden Republik Indonesia Nomor 41 Tahun 1996 Tentang Kawasan Industri [Presidential Decree No. 41/1996 on Industrial Estates]. Jakarta: Pemerintah Republik Indonesia.

Pt Jababeka 2010. *Annual Report: Ready to Capitalize, Ready to Grow*. Jakarta: Pt Jababeka.

Roy, A. and Ong, A. 2011. *Worlding Cities: Asian Experiments and the Art of Being Global*. Chichester: John Wiley & Sons.

Shatkin, G. 2008. "The city and the bottom line: urban megaprojects and the privatization of planning in Southeast Asia." *Environment and Planning A* 40: 383.

Stone, C. N. 1993. "Urban regimes and the capacity to govern: a political economy approach." *Journal of Urban Affairs* 15: 1–28.

Winarso, H. and Firman, T. 2002. "Residential land development in Jabotabek, Indonesia: triggering economic crisis?" *Habitat International* 26: 487–506.

9

MEGA-SCALE SUSTAINABILITY

The relational production of a new Lusaka

Matthew Lane

Introduction

Recent developments in the approach taken to planning in Lusaka, the capital city of Zambia, tie in closely with debates about city-making as part of urban policy networks. These universalise approaches to sustainability in the face of rapidly conflating social, economic and environmental concerns. Today, Zambia is recognised as one of the most urbanized countries in Africa with more than 40 per cent of its population classed as city dwellers (UN Habitat 2007). Created in 1931 to form a new administrative capital city of then Northern Rhodesia, Lusaka was established in a central location with good communication links to the rest of the nation. Influenced heavily by Ebenezer Howard's garden city plan, and framed by the controlling, power-driven ideology of the colonial project, the new capital was to be 'generous and spacious' in style and designed for a population of 500,000 (Mulenga 2003, Njoh 2009). The planning system subsequently failed to keep pace with demand for land during post-independence urbanization as Zambia embraced periods of structural adjustment. As a result, Lusaka is seen to have suffered heavily from attendant effects such as overcrowding and congestion, insufficiently serviced housing areas, rapid economic decline, urban poverty, unplanned settlements and poor living conditions (UN Habitat 2007). A realisation of the extent of these problems is at the centre of Lusaka's move to establish a brand new urban master plan for the future, a plan to urbanize the city-region. The production of this master plan provides the focus of the chapter.

In this chapter, I pursue the argument that in a global urban age, the process of urban planning, not least the planning of mega-projects, is the work of a diverse set of actors and complex power relations that extend across space and time. These ensure it is not only the comprehensive approach to planning that makes the project 'mega', but the scales through which the project is itself produced. Capitalist agendas, regional power structures and the institutional dimensions of planning seek to guide and control developments in various ways (Watson 2013). It is important therefore to understand the terms of engagement between the forces of globalisation, the practices of local planning actors, and the tactics and strategies aimed at both countering and engaging with these (Shatkin 2011). The chapter will be used as a means of providing an empirical contribution to the core themes of the book in relation to the 'fast' production of urban master plans, and to raise questions about how these plans are constructed. A particular focus will be on the nature of the implicit relations of power that work to 'assemble' both the plans and their institutional landscapes of construction. By introducing the master plan, its background and design principles, followed by a conceptual interrogation, the chapter will discuss a) the universalised nature of sustainable design discourse; b) the situating of government agency; and c) the legitimation tactics of the post-colonial state. In doing so, the chapter contributes to a re-theorisation of city-making in the global south through the ways it seeks to engage with discourses around sustainable development and represent them in master plans.

Lusaka's 2030 vision master plan

Presented as a comprehensive means of tackling Lusaka's burgeoning population and associated problems, the 2030 vision master plan has now been ratified and officially adopted as the long-term agenda for the city's development.[1] Like many of its counterparts in cities across the world, the plan bases its approach to a more sustainable future on principles of redesign that look to make a dramatic change to the overall fabric and structure of the city. Simultaneously targeting both the centre and the periphery, the plans integrate the spatial problems of land use and transportation, city and hinterland, by identifying the relationships between these two sets of issues as central to existing problems.

Labelled the '2030 vision' so as to cement its long-term commitment, the project is, at least on the face of it, primarily the product of the Japanese International Cooperation Agency (JICA) who worked alongside local partners to construct a plan that would solve the city's core problems of land use fragmentation, congestion and infrastructure degradation. Acting on the authority of Zambia's Ministry of Local Government and Housing (MLGH), as part of a long-term cooperation agreement between the respective governments, JICA was charged with carrying out an assessment of the current situation in Lusaka. As part of the study process, they identified what were deemed to be the core issues preventing sustainable development and attributed these to a threefold causality of: an old and spatially limited existent urban design; inadequate control of developments on the part of local authorities; and the burgeoning spread of the city beyond designated areas. As a means of addressing these concerns, JICA concluded that the new master plan must focus upon a new spatial structure for existing and future populations that included: i) desirable urban form for effective and sustainable control of sprawl and management of traffic; ii) appropriate density settlement for effective use of land; and iii) an attractive urban environment for competitive function as a capital city. These goals ultimately translated into the requirements of the overall plan.

To achieve the desired goals, JICA decided to frame the new design as a 'new urban expansion concept' and focused primarily on the spatial issues the city faces. This was rationalised as part of a required cohesive model, rather than tackling 'individual issues' such as 'residential neighbourhoods with inadequate housing' or a 'heavily used road'. The following principles are at the core of the plan:

Inner city area: Lusaka City territory

- Well-controlled dense settlement
- Efficient land use with adequate density distribution
- Dual-core central business district (CBD) connected by public transportation and pedestrian network
- Dual ring road construction
- Controlled urban growth by urbanization promotion area (within rings).

Satellite cities: adjoining three-district territory

- Self-sustained cities with dense settlement
- Planned settlement with adequate infrastructure.

While much of the focus is on the existing land use and transportation infrastructure within the city, the planned creation of satellite towns is deployed in tandem with this to remove congestion from the city centre and free up land for development, ultimately highlighting the intention to engage with spatial concerns in a macro way. A complementary initiative, in which large plots of residential land within formerly 'European' areas of the city are being bought from private owners by the government, not only serves to free up further land for development but also highlights how the authorities are looking to disengage with colonial ideals and practices. Addressing the issue of congestion through city-wide population redistribution clearly evidences the way in which authorities have framed the issue within the city as a whole and how their solution to the problem targets an equally comprehensive temporality. The incorporation of new satellite towns into the plan, however, means the project is not merely a city-wide initiative, but one that involves surrounding regions and districts. While this is somewhat necessitated by the existing spread of informal compounds into these districts, the desires of the government to diversify and expand the economic function of the city further justify the decision to focus growth in these new areas.

It is clear and widely recognised that land use and transportation form the central pillars of a spatial development master plan that targets the problem of sprawl. Through both the goals listed above and the more detailed aspects of the plan, concepts of sustainable planning and design become embedded into the vision for a 'bigger' Lusaka, bringing the city in line with the discursive paradigm currently governing the planning profession. These ideas and policies, however, have entered this particular plan through very specific channels. The physical agency of the Japanese consultants and their connections to the global pool of planning knowledge and resources is of course pivotal and something that will be explored in detail here. However, there is also the presence of more local and regional agency in the form of the fortunes of Zambia's closest neighbours to the south and the powerful perceptions of South Africa as a nation that has developed its cities somewhat more 'successfully' than Zambia.

The plan is currently in the early stages of implementation and while its spatial and material nature is interesting, providing an insight into the processes involved in reaching this stage is the primary remit of this short chapter. It focuses on the actors involved in producing the final plan and how it was constructed against the background of Lusaka's colonial history as well as other ongoing projects in the city. As will be demonstrated, the politics and processes that led to the production of the master plan reveal a great deal of insight into how a sustainable solution to solving the city's existing infrastructure problems can also be used as a vehicle for promoting and rationalising economic growth. Such goals rely on the ability of the plans to place Lusaka more firmly on the regional and global map while also engaging with the institutional and spatial impact of colonialism.

Relational production of scale in 21st-century master planning

As referenced in cases throughout this book, master plans provide an interventionist and comprehensive means of solving the crisis of urbanization and its sustainability concerns. It therefore goes some way to legitimising itself, in the view of the local authorities, on the grounds of the future that it proposes to deliver for the city. Why is this legitimisation possible? What is it about urban master plans such as these that gives them the seal of approval and appeals to audiences both near and far? These questions are further problematised by a tension that exists between acknowledgement of context-related diversity and the desire to produce normative theoretical positions which can be of generalised use to planners in practice (Watson 2003). As a result, both the problem and solution can be seen as constructed in extra-local settings, framed by the discourse of sustainability. Existing as a 'context-independent' interpretation of the planning role, political, social and economic rationales exist for the justification of master plans as a solution to the urban crisis (Flyvbjerg 2007). Politically motivated to be seen as addressing the most important, widespread issues, the state's decision to carry out large-scale projects such as infrastructure and self-sufficient satellite cities through the master plan aims to appeal to the economic investor as well as socially fractured communities.

Connections to a global discourse around sustainability through urban design have ensured that for a city like Lusaka, particular approaches to urban development are accepted as legitimate solutions to the crisis of urbanization and sustainability while others are rendered inappropriate. This results from wider 'global' interpretations of what is sustainable as certain ideas get moved from the realm of 'best practice' to become normalised practices off the back of their enhanced reputation as successful initiatives elsewhere. In this sense the discussion that this chapter wishes to build on is rooted in conceptions of mobile policy and practice (McCann and Ward 2011). A body of work that has itself built on more traditional political science literature on policy 'transfer', the concept of mobile policies has come to the fore in urban studies in recent years with the publication of texts such as McCann and Ward's *Mobile Urbanism* cited above. Identified by Jane M. Jacobs (2012) as one of the primary strands of 'relational' thinking in urban geography, policy mobility research seeks to chart the inter-city, multi-scale movement of policies and their architects in a wide range of urban sectors, such as business improvement districts, drugs policy, infectious disease control and urban governance tactics. Taking a broad political economy approach (Jacobs 2012), what sets this work apart from the more political 'transfer' literature is its desire to uncover the complex relations between policy sites and actors and the historically conditioned relations that shape the direction and materialisation of flows in the neoliberal era. Moving quickly through both time and space, these policy ideas, as techno-managerial projects of expertise and knowledge spreading, involve actors beyond the political who work in both locations of export and reception to frame the conditions of mobility. As a result, these mobilities are often part of more complex rationalisation processes with numerous motives (McFarlane 2006).

In relating this lens to the mega-urbanization of Lusaka, I wish to go a step further to look at how conceptions of sustainable design practice are bound up within wider discourse and complex power relations that serve to blur the boundary between the transfer of what Stone (2004) defined as 'soft' norms and 'hard' policies. While a number of cities, well embedded into the global circulation of ideas, may look to harness and learn from one another through the use of similar policies in areas such as those listed above, other cities may seek to take advantage of the latest trends and ideas with regards to a particular urban issue. In such cases, a methodological toolkit focused on 'following the policy' and making connections between particular cities and regions is insufficient. Instead, a more deconstructive interpretation, focused on conceptions of power and knowledge, is proposed as useful. At the core of the need to reinvigorate the debate between policy transfer/mobility and the looser or 'softer' globalisation of norms is the point made by Datta (2012) on the way in which the incredibly vague definition of sustainable development (as defined by the United Nations) has been translated and politicised in different contexts. It is upon this 'discourse'[2] of sustainable urban design that action is being taken when creating a new vision for a sustainable future. Indeed, as Watson argues,

> This conflict between the rationalities of governing and administration, and rationalities of survival (of those who are poor and marginalised), offers one way of understanding why, so often, sophisticated and 'best practice' planning and policy interventions have unintended outcomes (which is not to deny that other less explicit intentions may be driving these interventions).
>
> (Watson 2009, 2272)

Much of the discussion of best practice transfer seems to focus on cases where the mobility takes place between two mutually accepting cities or regions with the explicit intention to share ideas around a particular policy theme. When such a relationship is not present, the form of transfer takes on characteristics more suited to Stone's concept of softer 'norms', and while, according to Tomlinson (2002), this declining confidence over what constitutes the 'best practice' policy increases the ability of local actors to manipulate and adapt the specifics, it maintains its hold as a vehicle for the global, neoliberal agenda. This is perhaps particularly the case in cities of the global south, like Lusaka, where local authorities are keen to harness the latest ideas that litter the terrain of sustainable development but engagement with them does not reach the extent of a bilateral initiative with an 'exporting' city or region. As a result, a recognition of the locally specific cultural and institutional dimensions bound up within the policy does not figure (Gertler 2003). In view of this discussion, what is required is a means of integrating and assimilating the multiple discursive inputs and more material ideas of practice into a theoretical conception of master plan production.

As well as physical relations between actors located near and far, building on the above discussion around 'discourses' of sustainability, the role played by implicit 'global' agendas must not be neglected. As a means of doing this (while also maintaining an interest in the more explicit mobilisation of ideas between regions), the use of the *assemblage* concept with reference to the urban is particularly useful. While the term has been used in several different manifestations within urban studies alone (see McFarlane 2011a for a discussion of these), when applied specifically to urban planning, the concept of 'assemblage' promotes an immediate conception of a redefinition of the socio-spatial meaning of the term. It involves the composition of diverse elements and near–far, structure–agency relations (Anderson and McFarlane 2011). The concept allows us to not only integrate, theoretically, the expanding number and diversity of actors involved in a particular project (McCann and Ward 2011, xvi), but also to think about the more implicit roles of material and non-material agents within sustainable development discourse.

Unlike in the past for cities such as Lusaka where a core set of engineers carried out the task of urban design, the role of 'planning' today exists in a state of transition and contestation over its nature in terms of both *what* it consists of and *who* is responsible for partaking in it. Broadening our understanding of the roles played by actors in a globalised field of urban planning can take us beyond the obvious political functionaries of the state, to include networks of consultants and theorists as well as city-specific, territorialised agents such as NGOs and informal community leaders (Robinson 2005; 2011). Often based in extra-local settings, but having to engage with peoples and institutions on the ground in the city, the assemblage concept acts as a container for the various forms of travel, translation, struggle and connection that subsequently constitute urban development projects (McCann and Ward 2011). Despite this perhaps more recent proliferation of actors and agents involved in planning, there is of course still an interesting parallel on some levels between this and the colonial approach through the importation of planning principles from elsewhere and also, in Lusaka's case, in the naming and defining of the master plan as a 'garden city' plan.

It is within this expansion of how planning is defined that the assemblage concept has found its usefulness in attempting to understand the complex mix of knowledge and practice drawn from multiple scales, both spatial and temporal. In relation to the contemporary approach to sustainable planning and design, McFarlane's (2011) assemblage term is useful for breaking down the dichotomies between global discourses and local interpretations as well as between the material solutions in the form of explicit best practice policy adoption and the non-material in the form of framing, imaginaries and identity constructions. It is upon this point that I wish to build by returning to the role of normalised discourse in facilitating the design and construction of urban master plan projects. An important dimension of this is the role of *agency*, defined as the 'emergent capacity of assemblages' (McFarlane 2011b, 215). This agency also takes on a material nature both in terms of the existing situation on the ground in cities and the material manifestations of sustainability discourse in the form of media and policy documentation.

McFarlane's (2011, 206) discussion of assemblages is also useful for the way in which it frames the assemblages interpretation of the city as an entity through its interest in 'emergence and process and in multiple temporalities and possibilities . . . rather than as resultant formations'. In this particular sense he is referring to the city in general, but in the context of this chapter, it can be related to a master plan for the future of Lusaka's urbanization. Through its vision for the future, a city's own identity in the present is constructed. In this vein, assemblage thinking encourages us to focus more on the construction of the vision itself and trace it back to its inception as the product of particular discourses of sustainability, rather than on the finalised, holistic plan. It is through these agency roles that discourses of sustainability are acted upon in the transitional process of dis-course–action–materialisation, pertinent here to the construction of master plan mega-projects (Jager and Maier 2010).

As well as connections in geographical space, temporal connections between the cities past and present can also be highlighted through the plan's role as a vehicle to mobilise the city away from its contested his-tory. Implemented as blueprint city master plans, planning in the colonial era – the legacy of which remains for cities like Lusaka – was idealised as being designed to accomplish public welfare goals including sound archi-tectural standards and promotion of the efficiency and effectiveness of the built environment. However, more critical analysis such as that carried out by Njoh (2009) revealed that the real intention of such approaches was 'a meticulous agenda of the colonial powers to harness and reinforce their grip on the territories in Africa' (306). Compounded by rapid urbanization levels, which in itself is amplified by the fact that colonial planners failed to plan adequately for the large urban population of Africans that the new city eventually attracted (Myers 2003), the colonial legacy has translated into the set of core development problems mentioned above.

Moving forward to the current situation and the role played by the independent, post-colonial state in facilitating and subsequently imple-menting master plan mega-projects, Segbers et al. (2007) attribute the fact that cities exist as a plurality of actors and institutions that constitute a new 'patchwork politics' in place of the state as *the* key actor to develop-ments in the global economy. Put in different, more explicitly economic terms, the prospect of new cities, planned at grand scales, require enor-mous amounts of capital that is not at the disposal of the post-colonial state (Datta 2012). It is this fact that resulted in Choplin and Franck asking the question, 'Is the capital city up for sale?' (2010) given the valuable entry point that capital cities in the global south provide for privatisation through the influence of a new global class of actors. Keen to harness new ways of attracting investment, and integrating themselves into the global economy following post-1990s liberalisation, new master plan projects provide the perfect vehicle for states to develop new *governance* forms in place of redun-dant *government* structures (Pain 2010, emphasis in original). What this post-colonial melange of inherited colonial institutions and contemporary liberalised integration into the global market provides, from a theoretical point of view, is a great deal of potentiality that needs to be explored. In line with McFarlane (2011), 'potentiality' in this sense refers to 'the capacity of

events to disrupt patterns, generate new encounters, and invent new connections . . . in the form of blueprints, models, dreams or hope for a better
city' (209).

Designing a sustainable future for Lusaka

Examining the plan itself, Lusaka's 2030 vision project is ambitious (or
'mega') not only in terms of its spatial coverage but also in the aspects of
sustainability it wishes to engage with (i.e. all three of the economic, social
and environmental pillars). While often collectively discussed under the
banner of 'sustainability', in the context of a city like Lusaka, the different
branches of this sustainability discourse provide very different issues to
be dealt with. As well as creating an attractive investment environment to
encourage growth and deal with the myriad of social problems, the local
authority's desire to move away from the legacy of the colonial master
plan is seen as a further discursive element that contributes to the direction of the plans. This, however, is a particularly important narrative to
be pursued given the fact that the master plan approach to planning that
lies at the root of Lusaka's infrastructure problems is being revisited and
even the 'garden city' name recycled as part of the new vision. This can
perhaps be put down to the sustainable connotations that the term 'garden
city' invokes, emphasising the power of certain elements of sustainable
cityness.

The environmental strand of sustainability, being particularly powerful
in the contemporary era at both a policy and money-making level, is given
fresh impetus through the rebranding of the garden city concept. As visually appealing on paper as it sounds in principle, a well-designed garden
city plan has the potential to attract investment, particularly when some
of the constituent parts are tailored towards building on existing growth.
For example, it is clear that a number of the policy ideas being used in the
plan have been moulded around particular areas of the city where there
is either current economic growth or the potential for future expansion.
The is evidenced by the promotion of a dual-core CBD that moves economic activity away from the old central areas and clusters it in new regions
with greater potential – for example, along the main road axis leading to
the international airport, or within districts currently occupied by large
shopping malls. The sustainability discourse in this sense can be seen to
facilitate and legitimise certain neoliberal agendas. In some ways, then, the
environmentally grounded sustainability drive and the economic growth
that is correlated with the idea of achieving economic sustainability can be
seen to complement one another in legitimising the vision being presented.
However, questions could be raised over how the social dimension to sustainability and the need for a more equitable society as part of the complete
vision fits into such a mind-set.

The thought of a well-planned garden city stands up particularly well given the current situation of uncontrolled sprawl beyond planned areas, and poorly organised administrative functions in the city government. Proceeding to act upon the different discourses that define Lusaka's ongoing infrastructure problems, JICA, under the authority of the MLGH, set out to identify a number of potential comprehensive design models for the sustainable master plan. Despite being singled out as one of the root causes of Lusaka's current infrastructure and land use problems, the colonial master plan's construction as a 'garden city' remains an appealing concept, the reasons for which (beyond its environmentally sustainable connotations) remain unclear. This narrative of a reorientation and rebranding of certain aspects is maintained in the decision to describe the new plan as a 'new urban expansion concept'. Within this mode of thought, the problem is not orientated on the existent structure per se, but on the inability of both it and the post-independence authorities to adequately control urban growth. While urban 'sprawl' contains numerous negative connotations, urban 'expansion' is deemed a legitimate and sustainable means of constructing a future plan.

What both of these aspects perhaps illustrate is the importance placed by the Japanese on drawing attention to another reason for the existent problems; a discourse of institutional incompetence. In doing so, their role as the expert consultants is further rationalised, emphasising the values they bring to the table over any previous attempts to re-structure the city.

> It was time to come up with something big . . . something that could tackle the problems the city faces on a large scale. Other plans have failed in the past but through the help of JICA we have the plan we need; in line with the government's vision.
>
> (Interview, senior city planner, Lusaka)

The first stage of the process in terms of constructing the new master plan involved consultations between JICA and the local authorities and subsequently, a number of field surveys and discussions with urban 'stakeholders' as a means of reifying the city's spatial concerns. To the city council and local ministries, knowledge is everything when it comes to creating a city plan that is effective at both solving problems and taking Lusaka as a city to a new level of status within the region and indeed the world. The fact that they turned to JICA, a foreign organisation, to provide them with this knowledge and employ it in identifying stakeholders and carrying out subsequent analysis of the findings further emphasises the context-independent way that authorities in Lusaka view planning practice and the knowledge at its foundations. In many ways it seems the 2030 vision needed to have a mark of the 'expert' as much as it did of the 'global'.

We needed to draw on knowledge in fields of urban planning and development from elsewhere so we approached them because they are cooperating partners and have fields of assistance in urban sectors specifically. When we approached they were willing to help us and provided both the technical experts and the capacity to undertake the process.

(Interview, MLGH planning officer, Lusaka)

The action of seeking an external organisation to undertake the task of studying Lusaka's issues is rooted in a more embedded and long-term discourse of inadequacy and incapability on the part of the local authorities, established prior to the undertaking of this project. As acknowledged by JICA in their reports, the UN-initiated 'Zambia: Lusaka Urban Sector Profile' (UN HABITAT 2007) under the wider 'Rapid Urban Sector Profiling for Sustainability' initiative funded by the European Commission and governments of Italy, the Netherlands, Finland and Belgium serves to frame the institutional issues that exist in Lusaka within which the 2030 vision must build and of which a capacity to manage planning issues and lack of qualified personnel were raised as primary concerns. Discursively embedded within both the desires of local authorities to produce a recognised sustainability master plan and the donor-based support mind-set of the extra-local actors, the project is established against a background of ineptitude that renders it impossible for such a plan to be constructed without the input of foreign expertise. The following extract from the project documentation illustrates this point:

The capacity development issues in terms of city management administration are: i) strengthening development control; ii) improving land management mechanism for better living environment; iii) infrastructure project implementation; and iv) improving public administration . . . . Inspection and enforcement are also based on old regulations under limited enforcement capacity. Moreover, fundamental enhancement is primarily necessary to correct ineffective office works, incapable staffs with insufficient office skills, etc.

(Final plan documentation, JICA, 2009)

The above extract makes several references not only to the outdated planning regulations and lack of resources but also to the ineptitude of the staff themselves, and is situated within an extensive part of the master plan documentation that is devoted to addressing such required reforms and changes. This exists to the extent that the concerns over sustainability in Lusaka are as much institutional in nature as they are spatial or material, which ultimately serves to render the existing city planning authorities as much subjects of the plan as they are primary stakeholders in producing it. This view of the issues 'in house' served to rationalise the consultancy team as the primary leaders of the project, something accepted by the local planners because of their belief that while the 'financial help [from JICA] was important . . . the knowledge they could bring was of most importance in solving the problems in the city' (interview, city planner).

At a high level of city governance, there exists a clear agenda to cement the legacy of the 2030 vision not only through its physical characteristics, but also through ensuring the ongoing local authorities have the capacity, and crucially, 'knowledge' to maintain it. To do this, during the design and production phase, JICA had a number of individuals located within the Lusaka city council offices working alongside the local planners in producing the plans. As part of this process, however, at a lower level the city council with its limited resources must lay out plans to deal with the ongoing issues on the ground in Lusaka as the city undergoes this design transition. Invariably taking place in peri-urban areas, largely made up of informal, unplanned settlements, it is a branch of the city council known as 'Ward Development Committees' (WDCs) that are given the task of managing them. In working on a master plan and overall planning framework for the city in the long term under the premise that this provides the platform required for benefits to filter down to community level city-wide, in the shorter term alternative means are required. The council therefore requires, or in the words of the director of operations of the NGO Habitat for Humanity, 'expects' different players to take on certain institutional roles in the plan (interview, Habitat For Humanity, Lusaka. This perhaps illustrates the way in which the discourse of poor official capabilities has filtered out and become an institutionalised aspect of the problems faced by cities in the global south. The reliance on the role played by NGOs and their accepted contribution to the planning task in Lusaka is deeply embedded within the fabric of the city.

The borrowing of concepts from elsewhere and integrating them into the plan is a key part of ensuring what is produced is ultimately. Establishing what is required to make it a successful plan is not based solely on Lusaka as an independent case, but given its location, on its regional integration into the Southern African Development Community (SADC) economic region. This geographical determinism extends beyond the formalised organisations to include an agency that, as discussed earlier, is prescribed to a non-material discourse that exists as part of an international agenda-driven developmentalist framework and mind-set. Highlighting the agency of Lusaka's location in Southern Africa and particularly its proximity to South Africa, the following quotes, from interviews with a member of the Lusaka city council and following this from a JICA programme officer, explain the mutual thought process involved in determining the suitability and potential of development models.

> Of course when we started this concept we had some benchmark that said as much as we want this and that, can it be effectively applied here in Lusaka? Look at South Africa, they have a high population, their culture is this and this. Even the Japanese when they came are looking at what is happening in Africa, at what rate we are moving in terms of development to see what cities we can take to model ourselves on and come up with these designs. So we looked at South Africa. Cities in other nations such as Gaborone were also looked at but South Africa provides the best blueprint.
>
> (Interview, city planner, Lusaka, May 2012)

> It is important to learn from the way concepts have worked in cities in other countries. Throughout the creation of the plan and going forward, the way South Africa has developed its cities and the approaches they use have been assessed to see if they are applicable for integration into the plan for Lusaka.
>
> (Interview, JICA study team, Lusaka, April 2012)

Even though the discourses around sustainable urban forms may be global in nature, the use of certain ideas and models needs to be rationalised somehow given the divisive effects of developmentalism that has separated cities and regions of the world and led to a certain incommensurability between them. Using existing approaches and benchmarks set by other cities and counties on the African continent to channel the connection between Lusaka and the wider world is one way of doing this. While South Africa has clearly become an established model for Lusaka, the inclusion of Gaborone in the sentiments of the officials is particularly striking for the way in which it highlights the mind-set of following a particular pre-defined path, given the fact Gaborone and Lusaka are very different cities in the material sense of both form and size.

The influence of a number of factors, both in the shape of specific actors and in the form of development and sustainability discourses, are evident in the production of Lusaka's 2030 vision master plan, what is less explicit is the role played by the state in facilitating or contesting the finalised master plan for the city. What is clear, however, is the position taken by the authorities with regards to the existing situation and the need that existed for 'something' to be done.

> It is not only coming from us, we have seen the people, and the investors engaging and promoting the idea of a new plan because they are tired of the current system that is there. It is very cumbersome,

preventing development in the city. Something different is what is sought after and what the people want to see.

(interview, city planner, Lusaka)

While Lusaka's establishment under control and power-driven planning rationales may indeed serve to complicate the means by which sustainable master plan designs are grounded spatially, materially and institutionally it also makes the task, from the view of the city council, very clear: to acquire a sense of newness and distance themselves from previous planning approaches. In achieving this, however, the state must deal with what is described by Low (1996) as the 'politics of modernization' which involves the integration of the juxtaposed concept of master planning and the realities of life for the vast majority of the urban population who are unlikely to be even remotely engaged with the colonial history, but rather situated very much within the struggles of the present. In this vein, the terminology reflects much of the mentality of that used by the colonial planners when dividing and segregating the urban community.

Developmentalist urban discourses have positioned developing world cities as being less 'modern' than those in North America and Europe and lagging behind universal standards of development. Such discourses, as has been discussed here, perforate far beyond academic and media discussions and become embedded within the mind-set of those on the ground in such cities. This means that the issue of unsustainable growth and its associated problems becomes one that is not only spatial and material, but also institutional through the capacity that local authorities are deemed to have to solve such issues. As a result, the planning authorities in Lusaka wish to not simply tackle the issues the city faces but to use the new master plan as a vehicle for growth and development – as a strategy of mega-urbanization. The use of South Africa (rightly or wrongly) to justify and legitimise approaches to development on the part of the authorities as they seek to 'learn from its development trend' (interview, city planner, Lusaka) shows the power of this discourse in defining where the city is going. Having negotiated the ordeals of structural adjustment and privatisation that characterised the past two decades in Zambia as part of the wider motions towards political decentralisation and liberalisation that opened cities up to the processes of globalisation (Segbers, Raiser, and Volkmann 2007), the government sees this as the time to facilitate the country's direction towards regional integration. Lusaka will no longer be a mere capital city, but will be mandated to take the lead in the national economy in the future. Geographically endowed with proximity requirements such as markets, customers, suppliers and competitors, according to JICA's detailed study of the current situation, Lusaka exhibits potential in spearheading dynamic economic development as an extended urban region (JICA 2009).

However, as the city moves away from the design and ratification stage facilitated primarily by the Japanese, the responsibility is passed to the city council and the implementation stage . . .

> We have created the master plan which is now there but it is up to those authorities, such as the council, that can implement and conform to the more medium-term planning as they continue their search to fund areas of the project.
>
> (interview, JICA infrastructure projects leader,
Lusaka)

Time and the linkages between the local, regional and global situation hold the key to how Lusaka's vision is translated into an actual master plan on the ground. What is clear is that, from the point of view of those involved in the plan's production, capital investment is vital to this process of implementation at least in the short to medium term. This is because while a comprehensive plan that was well thought out with the greatest chance of long-term success was desired, a plan that was investable was absolutely essential. As with the 'Dubaisation' projects in Nouackchott and Khartoum that form the subject of Choplin and Franck's (2010) paper, the prospect of a master-planned mega-project paves the way from geopolitical marginality to an extreme form of modernity in the image of 'successful development'. In Lusaka's case this would mean harnessing new opportunities, facilitated by the new plan as a means of elevating itself in the context of southern African cities. What Lusaka does not have, however, are the financial resources at its disposal and what is therefore required, on the part of the state, is a form of agency that works to deliver an urban environment that is both investable and liveable at the same time. This would subsequently serve as the road to its greater recognition as a world class, modern and thriving capital city, symbolizing the whole country and functioning as a national and regional centre for policy, culture, science, technology, education, economy, and international trade.

Conclusion

This chapter has provided a brief insight into the processes involved in inducing mega-urbanization through master planning in a previously under-researched part of the urban world. In doing so, it brought this into conversation with concepts of relationality across both space–time and material–non-material dichotomies. Through an examination of Lusaka's 2030 vision, the analysis in the chapter argued that this form of city-making in the global south can be understood as a relational process bound up within the wider, normalised discourse of sustainable urban design and mega-urbanization of the region. Far from being a well-defined manifesto that acts as the starting point for the future, it is suggested that this plan can be most usefully imagined as the final product of complex relations between speed, time and space that frame and shape the city today and its future. A relationally produced interpretation of the problems and a discursively provided set of solutions ride on the back of not only a global discourse around urban designs but also embedded desires and ambitions of the local authority to emulate other cities and regions. In this vein it is about championing the need to take into account both the spatial relations at work in producing these plans and what they tell us about how the urban is conceptualised today, as well as the situating of existing issues in their historical context, as both the local authorities and the consultants they work with rationalise, legitimise and ground their decision making.

As part of more obvious, direct connections between utopian master-planning narratives and the cementing of long-term development visions, a number of legitimation tactics were used that reflect the embedded nature of discourse. It would seem that at the core of the city's plans is a desire to ensure lasting success by making the vision as marketable as possible so as to appeal to the realities of donor solicitation and private investment. As part of this legitimisation process, both the local authorities and their Japanese consultants identify certain requirements of a sustainability master plan that invoke an attractive environment, even going as far as to rebrand the original, colonial 'garden city' term as a vehicle of such rationality. There is also a clear belief on the part of the local authorities that local institutional capacity is insufficient to undertake the task of producing such a vision and that the required expertise is one that must come from extra-local, global scales. This need to ensure the resultant plans have a modernising capacity, never before experienced in post-independence Lusaka, reflects the temporal dimensions of relationality that correlate with the historical inefficiencies of local authorities with the attendant effects of an outdated colonial master plan. However, it is not only global relations at work; there is also the desire to learn from and emulate the successes of other cities in the southern African region as well as to integrate itself into more regional economic alliances.

What all of these aspects have in common is the impact and influence they had at different stages of the project, long before its ratification and adoption as the long-term agenda for the future. Conditioning the decision making over both the spatial and institutional dimension of master planning, numerous discourses and agendas become assembled into a politically charged and globally contingent contextual master plan. In making these points, among others, the chapter has drawn on a number of theoretical tools that it sees as useful for representing this means of imagining the relational construction of urban master planning in the global south. Given significant appraisal by

Jacobs (2012) in her discussion about existing theoretical paradigms that attempt to think about cities 'relationally', the assemblage approach has been at the heart of this toolkit. As has already been mentioned, it is not only a more diverse group of actors that is being assembled to produce the plan or indeed a group of concepts and ideas that make up the final product, but also the prescription of vital agency to a number of material and non-material discursive elements that direct and facilitate the whole production process. Focusing on the question of *how* multiple discursive and material agents serve to direct and frame both the requirement *for* and subsequent production *of* a new master plan in Lusaka, it is hoped that these arguments can stimulate and contribute towards further discussion over the way various projects of neoliberalism, developmentalism and modernisation (Roy 2011, 8) condition the processes by which sustainability becomes an institutionalised planning practice in different parts of the world.

In moving beyond explicit plans to examine the processes involved in piecing them together, we can trace the role of power through stages of discourse, action and materialisation. Referring to the varied history of critical urban theory, Colin McFarlane (2011) points to the vital task of exposing the power relations that drive urban forms and processes to become 'normalised', 'inevitable' and 'universal' (208). For Lusaka and other cities around the world engaging in processes of city building and indeed urban planning more generally, there exists a complex array of discourses and inter-scalar power structures that work to direct and facilitate the vision that is being sought. Because planning – when conceptualised in this way – can be seen to be as much about mediating different discourses and relationships as it is about designing and redesigning the physical urban landscape, it seems very likely that ultimate causality in the outcomes of these projects will be rooted in these very relations. While generalising beyond the specifics of this particular case is not the intention, the realities at play here can have wider impact on the way in which the political decision-making processes with regards to the direction of planning and the rationalisation of master plan projects is theorised in urban studies. It is hoped that this chapter can stimulate further work on planning in the global south and the circulation of policy and ideas that ground themselves in these locations through more discursively orientated lenses that attribute significant agency to a city's location in both time and space.

Acknowledgements

As well as all those who participated in the research, I would like to thank both James Faulconbridge (Lancaster University) and Saskia Vermeylen (University of Strathclyde) for their support during fieldwork and their helpful comments during the writing of this chapter.

Notes

1 The material presented in the chapter is based primarily on a period of fieldwork conducted by the author in April and May 2012. During fieldwork, twenty semi-structured interviews were conducted with key actors involved in the production of the master plan. This was followed up with subsequent analyses of policy documentation relating to the project itself and the wider discourse upon which the plan is based. All information related to the plan itself is extracted from the documentation provided by JICA and the Lusaka City Council. The plan can be accessed at http://open_jicareport.jica.go.jp/pdf/11932845_02.pdf
2 See Jager and Maier (2010) for a discussion of the way discourses exercise power in a society through the institutionalisation and regulation of ways of talking, thinking and acting.

References

Anderson, B. and McFarlane, C. 2011. "Assemblage and geography." *Area* 43(2): 124–127.

Choplin, A. and Franck, A. 2010. "A glimpse of Dubai in Khartoum and Nouakchott: prestige urban projects on the margins of the Arab world." *Built Environment* 36(2): 192–205.

Comaroff, J. and Comaroff, J.L. 2012. *Theory from the South: Or, How Euro-America Is Evolving Toward Africa*. Boulder, CO: Paradigm Publishers.

Datta, A. 2012. "India's ecocity? Environment, urbanisation, and mobility in the making of Lavasa." *Environment and Planning C: Government and Policy* 30(6): 982–996.

Flyvbjerg, B. 2007. "Policy and planning for large-infrastructure projects: problems, causes, cures." *Environment and Planning B: Planning and Design* 34(4): 578–597.

Gertler, M.S. 2003. "Tacit knowledge and the economic geography of context, or the undefinable tacitness of being (there)." *Journal of Economic Geography* 3(1): 75–99.

Jacobs, J.M. 2012. "Urban geographies I: still thinking cities relationally." *Progress in Human Geography* 36(3): 412–422.

Jager, S. and Maier, F. 2010. "Theoretical and methodological aspects of Foucauldian critical discourse analysis and dispositive analysis." In *Methods of Critical Discourse Analysis*, eds. P.R. Wodak and M. Meyer. London: Sage. 34–61.

Khirfan, L., Bessma, M. and Jaffer, Z. 2013. "Whose authority? Exporting Canadian urban planning expertise to Jordan and Abu Dhabi." *Geoforum* 50(10): 1–9.

Low, S.M. 1996. "The anthropology of cities: imagining and theorizing the city." *Annual Review of Anthropology* 25(1): 383–409.

McCann, E. and Ward, K. 2011. *Mobile Urbanism: Cities and Policymaking in the Global Age of Globalization and Community*. Minneapolis: University of Minnesota Press.

McFarlane, C. 2006. "Knowledge, learning and development: a post-rationalist approach." *Progress in Development Studies* 6(4): 287–305.

———. 2011a. "The city as assemblage: dwelling and urban space." *Environment and Planning D-Society & Space* 29(4): 649–671.

——. 2011b. "Assemblage and critical urbanism." *City* 15(2): 204–224.

Moore, S. 2013. "What's wrong with best practice? Questioning the typification of new urbanism." *Urban Studies* 50(11): 2371–2387.

Mulenga, C. 2003. "The Case of Lusaka, Zambia" *Lusaka, Zambia: Institute of Economic and Social Research*, University of Zambia

Myers, G.A. 2003. *Verandahs of Power: Colonialism and Space in Urban Africa.* New York: Syracuse University Press.

Njoh, A.J. 2009. "Urban planning as a tool of power and social control in colonial Africa." *Planning Perspectives* 24(3): 301–317.

Roy, A. 2011. "Urbanisms, worlding practices and the theory of planning." *Planning Theory* 10(1): 6–15.

Segbers, K., Raiser, S. and Volkmann, K. 2007. *The Making of Global City Regions: Johannesburg, Mumbai/Bombay, São Paulo, and Shanghai.* Baltimore: Johns Hopkins University Press.

Shatkin, G. 2011. "Coping with actually existing urbanisms: the real politics of planning in the global era." *Planning Theory* 10(1): 79–87.

Stone, D. 2004. "Transfer agents and global networks in the 'transnationalization' of policy." *Journal of European Public Policy* 11(3): 545–566.

Tomlinson, R. 2002. "International best practice, enabling frameworks and the policy process: a South African case study." *International Journal of Urban and Regional Research* 26(2): 377–378.

Watson, V. 2003. Conflicting rationalities: implications for planning theory and ethics. *Planning Theory & Practice* 4(4): 395–407.

——. 2009. "Seeing from the South: refocusing urban planning on the globe's central urban issues." *Urban Studies* 46(11): 2259–2275.

——. 2013. "Planning and the 'stubborn realities' of global south-east cities: some emerging ideas." *Planning Theory* 12(1): 81–100.

10

PLANNING NEW TOWNS IN THE PEOPLE'S REPUBLIC

The political dimensions of eco-city
images in China

Braulio Eduardo Morera

In a recent debate at the Barbican in London, Theodore Dounas – a professor of architecture in Xi'an – proposed that Chinese city making is a process in which Western planning practices are not an appropriate framework for analysis. His presentation highlighted that, contrary to Western reactions to new plans, the Chinese 'recognise the basic intent, that even colossal scales of planning are ultimately intended to improve the lives of everyone' (Self 2013, 104). This point, passionately discussed on the occasion by a few UK-based practitioners, highlights a key starting point for this chapter: our understanding of the Chinese planning process is quite limited. The 'colossal' urbanization in China – which for Western practitioners most times seems uncontrollable and exotic – can hardly be analysed using our existing practices as a lens for examination. As Wu (2012) suggests, the recent process of urban development in China has been dominated by three main stages: the fever for 'development zones' in the early 1990s; the 'global city' in the later 1990s; and the current enthusiasm for eco-cities. This enthusiasm, as Wu clarifies, comes from local governments – and municipal governments in particular (Wu 2012, 169).

Eco-cities are continually mentioned in the media and have become an important issue of local scholarly investigation in China, particularly in urban planning. Wu suggests eco-cities are often 'presented as a kind of new green paradise' by their designers and promoters (Wu 2012, 169), a common pattern across the imagery in most eco-cities worldwide. However, the process and practices associated with their planning and implementation still seem indecipherable for Western practitioners. This lack of understanding has raised a number of questions among Western academics and journalists on whether the fever for eco-cities is simply motivated by a desire to acquire the latest foreign urban model imported from the West.

Building new cities in China is a complex undertaking that goes beyond urban planning and architectural representation. Chinese eco-cities are not only the product of planners and their private clients, but also politicians and a number of other actors, inside and outside China. Several eco-city projects have included international 'green gurus' and large multidisciplinary teams. In many cases foreign investment and international politics become part of the mixture. For example, in a state visit to Shanghai in January 2008, the then Chancellor of the Exchequer in the UK, Gordon Brown, mentioned Dongtan eco-city and its images as an example of cooperation between the two countries. In his words, Dongtan showed the 'higher level of cooperation that now exists on one of the greatest issues that face the world, the environment' (Moore 2008). The design was also presented by China at the United Nations World Urban Forum as a world-class example of an eco-city. It was even mentioned in the US embassy cables released by WikiLeaks (Hesse 2010). Tianjin eco-city, on the other hand, has international politics as its origin. The initiative of implementing an eco-city came from the Singaporean Senior Minister Goh Chok Tong, who approached the Chinese Premier Wen Jiabao in April 2007 with the idea of jointly developing an eco-city. This initial approach was followed by the selection of Tianjin as the preferred location and a framework agreement signed by Singapore Prime Minister Lee Hsien and Wen Jiabao in November (Ng 2009).

Is this eco-city 'fever' in China any different from that taking place in other fast-developing geographies? What makes the Chinese case different, if different at all? In previous publications, Datta (2012) suggested that the legal and policy context is a critical approach to understand the implementation of eco-cities in India. Her analysis concludes that in Lavasa eco-city private developers have often taken advantage of the weaknesses of the environmental policies. Nevertheless, the case of Lavasa eco-city confirms that the Indian 'model' is one that 'makes claims to sustainability and is delivered through private investment' (Datta 2012, 992) – an approach that has already been unsuccessfully used in the United Kingdom. A privately driven model, however, seems to be unlikely for East Asia. Hyun Bang Shin (2011) suggests that the role of the state is more prominent in East Asian nations. In his words, states have created 'close ties with business interests and performed strong state intervention during investment decision-making' (Shin 2011, 50). How are Chinese eco-cities subject to strong state interventions? Is this affecting every aspect of the project – including their representation – or are there specific points on which the Chinese state focuses its actions?

The evidence presented in the following pages will suggest that China is leading a trend in East Asia. However, it would be too simplistic to suggest that there is one 'Asian' response to contemporary challenges associated with rapid urbanization. The two landmark eco-city projects explored in this chapter – Dongtan and Tianjin – provide evidence that China's current approach is complex and involves the role of both private and governmental actors in multiple arrangements. Importantly, the examples explored also show that the design and implementation of large-scale projects are dramatically influenced by Chinese-specific variables such as politics and local forms of communication associated with these. In summary, I propose that without acknowledging the 'political' as a force of urbanization, academics and others interested in China can hardly grasp the motivations for eco-city development and other forms of urbanization.

The design of eco-cities in China has been distinctively characterized by the abundance of images as a form of communication of the ideas. As in many other aspects of politics in China, images play a key role in depicting and promoting 'new' ideas. Grace Ng, a Singaporean journalist following the experience of Singaporean officials in Tianjin eco-city, clarifies that a key motivation for the local officials has been 'to build eye-catching projects with tangible results . . . that will earn them a promotion every three years' (Ng 2009). Understanding the content of these 'eye-catching' images is therefore a good way to appreciate what physical elements included in the eco-city model are of interest to those promoting and supporting the projects. Is there a perception that eco-cities have the capacity to boost land prices higher than other types of development, such as changes in government regulation, or can achieve greater returns than other types of economic instruments such as changes in the regulation? Or are these simply visions that better embrace the objectives and promises of a fast-urbanizing China?

The following pages explore the role of eco-cities as an arena for interaction between property developers; national, provincial and local-level planning officers; and environmental agendas such as ecological urbanization that are defined at ministerial level. This analysis aims to encounter the instances in which the design of an eco-city, both as a spatial production and a planning process, creates opportunities to reconcile the sustainability objectives recently promoted by the Communist Party of China (CPC) and the aspirations of politicians and the emerging middle class. Masterplanning images, on the one hand, are proposed as valuable evidence to understand key messages that are shared in visual format. In China, colours, figures and composition in urban landscapes can be considered as politically charged messages that are important to consider. While the 'design' aspects of eco-cities – morphologies, technological choices, imagery, and marketing practices – have followed common patterns, they are not simply the direct application of Western ideas. Western architecture is normally combined with strong messages of harmony, national pride and success in these eco-cities. Complementing this, the processes associated with the 'delivery' of these designs – financing, planning, implementation – are specific to China and other East Asian nations and are explicitly related to the politics of land development. To build the argument, the role of various forms of power, control and influence at various state levels is important to help us understand the complexities of urbanization in recent years in China.

Current eco-city developments in China

In 2012, the 18th Congress of the Communist Party of China (CPC) passed a modification to its constitution adding 'promoting ecological progress' as one of five ingredients to building socialism with Chinese character-istics (Xinghua 2012). This modification defines that one of the roles of the CPC is to 'lead the people in promoting socialist ecological progress'. According to the constitution, in addition to conservation actions, socialist ecological progress will be achieved by pursuing 'sound development that leads to increased production, affluence and a good ecosystem' (China. org.cn 2012). This explicit reference to principles of sustainability comes to crystallize the consistent discourse that China has been producing on sus-tainable development, in which eco-cities are flagship projects. Moreover, a white paper on climate change policies, created in 2011 by the Chinese State Council, declares that China's government 'has always attached great importance to climate change' (PRC State Council Information Office 2011). According to China's Vice-Environment Minister, Li Ganjie, 'more than 1,000 cities and counties have worked out blueprints and timetables to achieve eco-civilization which features harmonious relations between people and nature' (Xinhua 2011). The strong involvement of the govern-ment in all its levels suggests that the adoption of eco-cities is likely to be a matter of importance for the government. The inclusion of sustainability principles applied to urban development – and eco-cities themselves – in official policy documents can be understood as key evidence of their stra-tegic significance within the national environmental and political agenda. Eco-cities, as with many actions in China, can be understood as political objects which respond to current motivations of the regime. Importantly, the recent developments of eco-cities in different forms suggest that, in addition to the economic and political motivations that currently domi-nate Chinese urbanization (see Hsing 2010), some actors have a genuine interest in exploring the environmental benefits of using this model in the country.

Without clarifying what an eco-city is, the World Bank has suggested that, in practice, the analysis of eco-cities should include two main catego-ries: those that have been certified by the central government and those that have called themselves eco-cities by setting high standards of urban develop-ment (Baeumler et al. 2012). According to the Chinese government, only a city already qualified as a National Environment Protection Model City can officially become an eco-city. The large majority of projects that call them-selves an eco-city do not follow this procedure, thus a number of projects have been informally legitimized as eco-cities due to their high profile. These include 'the Sino-United Kingdom Dongtan Eco-City in Shanghai, Caofeidian International Eco-City in Tangshan, Sino-Swedish Wuxi Low Carbon Eco-City, Sino-Finland Mentougou Eco-Valley in Beijing, and the Sino-Singapore Tianjin Eco-City (SSTEC)' (Baeumler et al. 2012, 38).

The inconsistent approach in the identification of eco-cities has also been observed in a number of sources consulted in this research, including Chinese and Western sources. For example, the World Bank research team (this includes a mix of Chinese and non-Chinese researchers) already quoted identifies 11 official eco-cities according to the Ministry of Environmental Protection (MEP) as well as six unofficial (Baeumler et al. 2012). Other sources, such as the Eco-city Notes group – also a mix of Chinese and international researchers – identify 21 (Connelly et al. 2012), while the global survey published by the University of Westminster (Joss et al. 2011) identifies 25 eco-cities in China. The large majority of the eco-city initiatives recorded by the Eco-city Notes collective (Connelly et al. 2012) and Joss et al. (2011) correspond to projects controlled and promoted by municipal governments. Interestingly, the second largest number is associated with initiatives in which foreign governments such as Singapore, Taiwan or Germany cooperate and participate with the Chinese state at their various levels. Although these are widely publicized, private initiatives such as Dongtan constitute a minority in relation to the non-private initiatives. This, however, does not preclude investors from eagerly participating in these types of initiatives. Municipal and governmental initiatives are plagued with joint ventures and, subsequently, the promoters are constantly seeking to attract more investment. Similar to other types of land development in China, securing new investors is considered a major achievement and as such is widely publicized in the press. Arguably the main difference in the cases covered is that the international cooperation status of the investments seems to be able to attract a wider range of investors, who would prefer this type of development to the detriment of other conventional, 'non-sustainable' projects.

Three out of four private eco-city initiatives were initiated between 2004 and 2005 with dissimilar levels of success. These include Dongtan and Wanzhuang, both promoted by the Shanghai Industrial Investment Corporation (SIIC) in peri-urban locations of Shanghai and Beijing, respectively. Additionally, the Vanion Group based in Beijing, also with the technical input from Arup, started a project in that period. This initial stage also included projects where other arrangements were tested, such as Huangbaiyu (the cooperation between non-governmental organizations and a green 'guru'); Yangzhou (cooperation between GTZ – German International Development Agency, now called GIZ – and the municipal government); and municipal initiatives such as Rizhao and Chongming Island. Perhaps motivated by the apparent success of the private initiatives, the year 2006 shows an important shift in the leadership of this type of project towards municipal governments. The Ministry of Environmental Protection promoted 12 projects as designated eco-cities or national eco-garden cities. In the following years, 2007 to 2011, there was much cooperation between foreign governments – in particular between Singapore and Germany – and the Chinese central government (Tianjin, Suzhou, Qingdao); the Government of Singapore and the Jiangsu provincial government (Nanjing Jiangxinzhou Eco Hi-tech Island); as well as foreign governments (Taiwan, Sweden and France) with municipal governments. These waves of eco-city types arguably can be associated with a specific demand to test and experiment new types of development promoters, probably after witnessing the cancellation of the privately owned Dongtan Eco-city in 2006.

Converging representation

According to Ross Adams (2010), 'speaking of the *design* of [eco-city] projects is itself a convoluted task' (Adams 2010, 3). Adams highlights that eco-city design is plagued with a number of common design artifices such as eco-corridors and wind turbines. Moreover, he continues, green elements such as foliage and landscaping are used to hide the proposed buildings, creating a phantasmagorical figure-ground illusion that ultimately confuses the observer. By briefly describing Foster & Partners' images for an eco-city in Mexico, Adams gives an account of an awkward lack of architecture in the proposals for new development. Although in the case of Foster & Partners' work this position can be attributed to the need to create an atmosphere – a speciality of the architectural practice's rendering consultants, Vyonyx – in the case of Chinese eco-cities the generous use of trees and landscape elements is odd. Adams characterizes this tactic as a 'rhetorical inversion with regards to the inherent virtue of urban design' (Adams 2010, 6). While this opinion may be perfectly valid in rather dry areas where Foster's eco-cities have been proposed – Mexico City and Abu Dhabi, for example – it is important to highlight that in the case of China the use of many natural or built environments in public imagery has different connotations. Traditionally, watercourses, hills and birds directly relate to traditional messages in Chinese culture (Fu 2010). This theme, not discussed so far in the case of eco-cities in China, has been previously explored by the sinologist Lester Ross (1988) who identifies nature as one of the key elements in the creation of conservative propaganda in the early 1980s.

In the case of the visual production associated with eco-cities in China, although these images are not commissioned directly by the government, they can be chosen by the government to represent its politics. Moreover, because of large-scale and high-level investment, eco-cities will attract the attention of the media and subsequently they become projects of national, or regional interest. Examples such as the Beijing Olympics Park, Shanghai World Expo and eco-city projects such as Dongtan and Tianjin have been particularly successful in demonstrating in visual terms the philosophy of the CPC, and are thus being used to showcase Chinese progress.

In the last few decades this visual production gained a key role in demonstrating China's capacity to embrace its developmental and environmental challenges by creating strong domestic and foreign attention. Following a tradition of urban posters created in the 1980s and 1990s to promote the achievements of coastal cities, the recent images of low carbon cities, eco-cities and other green districts could be regarded as new posters of the regime which are used internally and outside the country to demonstrate China's effectiveness in coping with a deteriorating environment and rapid urbanization.

In the Chinese media, images of new development projects are common. They are usually published by local governments as a tool to attract investment and visitors by marrying ideas of urban sustainability and climate change adaptation with marketing practices. In the case of eco-cities, it is common that these images are explicitly focused on highlighting the interaction between human-made and natural features, as well as a certain sense of order only interrupted by special features – called 'nodes' in Chinese planning.

The Chinese planning process allows the strong influence of local officials and planning authorities in the communication of urban projects, even if these are created by foreign designers. A common practice is that high-profile proposals usually seem to reach implementation stage after the proposals have been substantially modified. In practical terms, the local design institutes take all project information and modify it according to the desires of clients, negotiations with foreign designers and their understanding of the local regulation. These practices have an impact on many elements of design such as changes in the overall layout of urban projects, their graphic representation or even types of buildings included in large-scale projects. In addition, the adoption of the 'philosophy of harmonious society' in 2004 was arguably the key event that allowed the validation of the 'eco-city image' as a governmental planning practice. This process of adoption of a green aesthetics that comprises natural elements and visual traditions is a significant process in understanding eco-cities as a visual representation of government politics. On the one hand, these images of eco-cities are surprisingly similar in their content but, on the other, the projects follow dissimilar planning routes. Thus, is the representation of cities through green aesthetics, in particular that of eco-cities, a product of architectural vision, a political bet or both?

Dongtan eco-city is perhaps one of the first cases in which the propagation of eco-city imagery reached a global scale. Although its design was not implemented in its original form, Dongtan represents one of the first widely publicized examples of eco-city design in China. While the validity of the heavy criticisms about the failure to implement the project is undisputed, its developer, the Municipality-owned Shanghai Industrial Investment

Figure 10.1 South canal in Dongtan

Source: Arup

Corporation (SIIC), kept on its website many of the images produced by the British engineering and design consultant Arup until very recently. The image above (Figure 10.1), developed around 2005 by Arup and Oaker, portrays a pedestrian promenade along the canal crossing the South Village of the project. In the foreground we can see abundant low vegetation and birds of different species by the canal where a rower seems to move through the water, causing little disturbance. On the right-hand side, there is a pedestrian promenade which is actively used by several people of different ages. The sharp and geometrical edges of the promenade as well as a line of small wind turbines create a strong contrast with the untamed vegetation surrounding the canal. In between the tree lines, quite uniformly distributed on both sides of the canal, we can also distinguish a few low-rise buildings that seem to keep the same scale as defined by the trees but which clearly establish a contrast with the vegetation because of their volume and dissimilar colours. In the background, an arched truss bridge defines the end of the perspective, again creating an exception within the abundant foliage contained in the image.

The use of contrasting elements such as foliage and geometrical promenades, or the canal with sinuous edges and an arched truss bridge, reveal the problem of the design of the so-called 'eco-city'. Although the union of natural and human-made elements is clearly the message that the image intends to convey, the different character, colour, texture and shapes of the natural and human-made elements in the picture expose visually the practical problem of designing a human settlement in direct contact with nature. The use of such elements, as in most architectural representations, responds to influences that go beyond the use of a 'style'. In this context, this peculiar use of elements is probably responding to a number of influences, which are likely to have their origin in both Western and local ideas.

Other eco-cities also showcase generous public spaces and extensive lawns as well as water features in their images. High-rise residential buildings appear as isolated, and quite modernist, elements normally in the background. What stands out in these images is that although the eco-cities in China are normally part of a dense development, the main element that is represented is a green space that is meant to look untouched and secluded from the residential areas. Once again, the use of elements traditionally seen as opposites such as high-tech urban infrastructure and pastoral landscapes is hardly a consequence of the preservation of the natural elements of the site.

Although most eco-city images haven't been commissioned directly by the government, these have been chosen by representatives of the state at the local or national levels to be used to represent their policies in certain events, such as in the visit of the former British Prime Minister Gordon Brown, visits of the Singapore Prime Minister Lee Hsien and an eco-cities conference in Shenzhen – all in 2008. Moreover, because of the large scale and high level of investment, eco-cities attract the attention of the media and subsequently become projects of national or regional interest. Similar to other major urban projects such as the Beijing Olympics Park and Shanghai World Expo, the unbuilt eco-city projects such as Dongtan or Tianjin have been particularly successful in demonstrating visually the philosophy of the current government and consequently have been used to showcase Chinese progress. As part of this trend, eco-cities represent a relevant example to analyse how the 'green aesthetics' of eco-city images can be instrumental to the government's need to represent its achievements at national and international level.

Dongtan and Tianjin provide evidence that shows why eco-cities are attractive to authoritarian regimes and how these regimes develop their own strategies for the implementation process. These images, however, also demonstrate a curious emphasis on unlikely contrasts of elements such as wild nature with generous amounts of high-technology devices. The use of visual elements of national pride (such as green landscapes in long perspectives) as well as traditional cultural icons (birds, peasants) shows an eclectic mix of objects that goes beyond the illustration of 'green paradises' (Wu 2012) or the greenery described by Adams (2010).

While eco-city images have achieved a substantial degree of dissemination, the visions set by these illustrations are far from achievable. Thus, are these images a visual fallacy, maybe perceived by the state officials as mythical imaginations of a future China – and thus a new type of propaganda – or are these idealistic images simply inspiring visions of the development proposals that Chinese cities should put in place? Arguably, either option is possible because eco-cities provide a flagship project able to illustrate world-class aspirations and a capacity to innovate for the whole country, that then become objects of desire and political control.

Piloting the concepts

Urban development that involves flagship projects respond to a process of planning that, until now, follows a conventional process of design and approvals supervised by the local government. It is therefore unusual that private developers adopt the eco-city agenda in major development projects. So far, in China only a handful of private projects (Changxing, Dongtan, Langfang, Wanzhuang and Yinggehai) have been released and declared to have eco-city aspirations in the last decade. Most of them have faced major problems for the approval of their land quotas, probably as a demonstration of the importance and conflicts of the state actors in this process. Despite the unfavourable context dominated by the more experienced local officials, these projects have not had problems in attracting the attention of foreign investors, designers and the press, becoming known not only in mainland China but also abroad.

Ambitions and expectations about Dongtan were high a few years ago. Following the signing of the contract for the design of the initial 650 hectares of the eco-city in early 2005, then Prime Minister Tony Blair launched the 'Dongtan macro-cooperation agreement' for the design of eco-cities in China between SIIC and Arup (Cao 2011, 488). According to SIIC's website, that 'was a hottest spot on November 9, 2005' (SIIC 2011). During the following year, at the same time in which the first reports on the troubles in developing William McDonough's Huangbaiyu emerged in the press, Dongtan was perceived as a success story. At this point, the Chinese government participated in a number of high-profile activities, including signing agreements and publicizing project images. The government's support was promptly withdrawn with the detention of Chen Liangyu, by then Mayor of Shanghai and leader of the local CPC in September 2006. Following this, the central government shifted its attention to a new project proposed by the Singaporean Prime minister, Tianjin eco-city.

Although they are created in an atmosphere of high enthusiasm and promises of change, eco-cities in China prove to be quite complex at the stage of implementation. While the flow of optimistic press releases, publications and conference papers on eco-cities in China seems to be constant since the early 2000s, news about the subsequent failure of many of these projects are also not unusual. A good example of this dichotomy is Dongtan. The Arup team, assisted by the Shanghai Urban Planning Design Institute (SUPDI), issued a 'control plan' application which, according to Peter Head, 'was given outline planning permission in the summer of 2006 and building work was planned the beginning of 2007' (Head 2006, 17).

However, only a few wind turbines were installed before the works stopped by that time.

According to Hong Cao, several hypotheses have been given in the blogosphere and newspapers for the cancellation of the project. He synthesizes them in the following groups: '(1) The foreign design company don't know China's national conditions and the practical needs of the local residents'; '(2) The dispute on the actual investment makes Dongtan eco-city lack sufficient and stable financial support so that the project has been stopped financially'; and '(3) the alternate of Shanghai party secretary has put off the process of the project' (Cao 2011, 488). These hypotheses bring us back to key questions on the application of sustainability principles in the Global South, such as: the appropriateness of foreign design and strategies; the economics of sustainable initiatives; and, perhaps more importantly, the social circumstances and implications. Was the project stopped due to a lack of technological research or planning expertise in China? Or was it simply in an economic dispute between the designers and the clients? Either of these concerns would have been too easy to resolve or negotiate considering the scale of the financial investments at stake. What is rather more significant is that the change in the positive mood surrounding the project team was dramatically disturbed by the detention of Chen Liangyu, the then Mayor of Shanghai and leader of the local CPC, in September 2006. From that point, positive press releases and the certain development of the project came to a halt.

According to SIIC's public information, SIIC is owned by the Shanghai Municipality. As in many examples of state capitalism in China, big economic conglomerates are owned by the government, and although they behave as private companies, they make use of preferential access to contracts and stock listings. On the other hand, according to reports of the *New York Times*, Mr Chen 'resisted central government demands to reduce speculative real estate investment and clamp down economic growth to prevent waste and overheating' (Brenhouse 2010), creating a power struggle between the central government and local party. As a consequence of this episode, widely covered by the international press but scarcely mentioned by Chinese planning scholars or commentators, SIIC did not carry on with the development of Dongtan. Although after these events both developers and designers attempted to maintain the project as one of the schemes for the 2010 Expo in Shanghai, Dongtan's participation remained under a cloud of uncertainty. However, Dongtan did not simply disappear from the media and the world of international politics.

Despite all these efforts to keep the project on course, the uncertainty remains. Professor Tao Kanghua from Shanghai Normal University suggests 'the problem is rather complicated, hard to describe' (Zhongjian 2008). He explains that after 2006, and 'mainly because of China's land policy tightening in recent years since the implementation of the central vertical management of land, approval of local land rights [were] gradually withdrawn' (Zhongjian 2008). In the case of Dongtan, land is considered as state-owned reclaimed agricultural land; hence, if converted into construction land, it is bound into the complex process of land conversion, approval of which is managed by the central government. In 2010, Peter Head, interviewed by Treehugger, said, 'implementation of the master-plan we produced has been postponed. As far as we are aware, this delay is indefinite and we don't know the reasons behind this'. Similarly, the county magistrate of Chongming, Zhao Qi, declared in a Shanghai government press conference in January 2010: 'There is still no timetable for the development of Dongtan eco-city'. He continued: 'because the ecological sensitivity, technical reservations and the sequence of the exploitation . . . there may be minor adjustments' (Xuhui 2010). These contradictions and ambiguity support the notion that eco-city implementation responds to a 'selective incorporation of ecological ideas' (Pow and Neo 2013) where the state remains in control. In practice, this process demonstrates itself as an inconsistent series of events that are very difficult to understand from the perspective of those who are not familiar with Chinese planning processes. Nowadays, almost 10 years later, the Dongtan site is under development but sources at Arup suggest they have no involvement in the current scheme.

'We can do it our way'

Tianjin Eco-city, however, has been a different story altogether. Jointly supported by the Chinese and Singaporean governments, Tianjin eco-city was developed in a 30-square-kilometre site and plans to provide housing and amenities for 352,000 residents. Although its planning objectives are quite similar to those of Dongtan or most eco-cities, its developer, the Sino-Singapore Tianjin Eco-City Investment and Development (SSTEC), emphasizes both commercial viability and civic values as fundamental principles of its sustainability (SSTEC 2010, 2). Arguably, this idea can be understood as an attempt to differentiate it from other eco-city projects such as Dongtan which were perceived as economic and political failures.

According to Stanley Yip, a Chinese planning practitioner, Tianjin shows that 'the planning for Eco-City has thus been elevated to the level of inter-governmental direct cooperation' (Yep 2008, 3). The consolidation of this cooperation not only in the design of a masterplan but also in the definition of a business strategy has proven successful in many aspects. For example, by September 2008 the project broke ground and it is currently confirmed that several international developers will invest in the project. However, this alignment of governmental powers does not seem to be exempt from problems associated with the bilateral nature of the planning and implementation process. In 2009, Grace Ng published in *The Straight Times* an article that compiled a series of interviews with members of staff working in Tianjin Eco-city. Among these, an interviewee mentioned that 'we have very strong high level government links – but not with officials at the lower and provincial levels' (Ng 2009). Similarly, the Chinese Ministry of Commerce (MofCOM) and Chinese Ministry of Foreign Trade have developed strategic associations with other foreign ministries, such as Singaporean (in three projects other than Tianjin eco-city); French (in three eco-city projects); and the German Government, the latter in two projects through its German Federal Ministry of Economics and Technology (BMWI) and its international development agency, GIZ. For eco-city developments, these connections are currently perceived as essential, the lack of which some commentators such as Cao (2011) have criticized in the case of Dongtan. The diversity of forms in which these relationships are established – Premier to prime minister, ministry to ministry, ministry to development agency – shows that, as argued above, the central government is keen to experiment with several types of development arrangements.

The state actors at national level have also taken an active role in the development of a sustainable language for the promotion of eco-cities. For example, drawing on the experience of Suzhou Industrial Park,[1] the National Government has taken part in the development of new towns. Currently, Tianjin eco-city has become the flagship project for the national government. Developed in partnership with the Singaporean Government and Tianjin City Government, Tianjin eco-city has become the key reference for other eco-city projects. Here the Chinese government is proposing the application of state-of-the-art principles of sustainability and low-carbon development to city design, even if they include principles of sustainability that, if fully developed, could challenge the political status quo. Although the developer entity, SSTEC, is formed equally between the Singaporean and Chinese governments, each seems to be driven by different motivations. As one of Grace Ng's Singaporean sources for her article in *The Straight Times* suggests, 'local officials appear to be no longer willing disciples of the foreign experts'. Moreover, Ng continues, the Chinese officials seem to be unwilling to apply principles of city development from abroad to their local needs. In fact, she quotes a Chinese official who openly stated in a project meeting, 'we have already learnt Western management techniques. Now, we can do it our way' (Ng 2009).

Besides the obvious questions on what the actual motivations are for the Singaporean cooperation, it is important to ask whether these views also extend to principles of sustainability. In this sense, are sustainability and eco-city implementation understood as a foreign influence, and if so, what are the local influences and parameters that local officials and politicians would understand as offering success? You-Tien Hsing (2012) proposes the main objective of any official involved in urban development is the increase of local value. In her words, 'property prices are used to measure the success of urban development, and are openly referenced by local leaders as primary political mandate' (Hsing 2012, 9).

The success of Tianjin – the first major eco-city in construction – has been widely publicized by several actors; for instance, at international diplomatic events, such as the visit of Singapore's Prime Minister in September 2012, and in Chinese planning publications such as the China Planning Review (Yang and Dong 2008). According to Yang Baojun and Dong Ke, two of its planners at the China Academy of Urban Planning and Design in Beijing, Tianjin's significance as an eco-city is mostly because of its very construction and delivery (Yang and Dong 2008, 33). This success, however, is not purely a technical achievement, but also a political one. Tianjin eco-city shows that the alignment of motivations of government at all levels is one of the key aspects for delivering a project – an ingredient missing in Dongtan eco-city. In the case of eco-cities, these motivations are influenced by the political conditions, but these deserve detailed analysis on a case-by-case basis because they are affected by different market and state actors depending on their location and their relationships.

Shaping eco-cities

While eco-city designs produced by international designers are well received and promoted by local authorities, these are usually adapted and transformed so that they comply with Chinese policies and, more importantly, state actors' expectations. In whatever stage of the planning process these projects are, they can become governmental flagships, even if the images of them change later in the development. The involvement and importance given to eco-cities by the local and central governments have raised questions in regard to the real motivations of the application of the most advanced principles of sustainability and low-carbon development to city design. While the motivations of Chinese officials are one of the variables used by Western scholars and commentators to criticize and judge the relevance of environmental measures – particularly eco-cities – this is an issue urban geographers are just beginning to grapple with. It is in this context that this chapter has provided a critical analysis to guide further research.

Dongtan and Tianjin show why eco-cities are attractive to authoritarian regimes, and how these regimes develop their own strategies to deliver and realize these on the ground. Arguably, this is possible because eco-cities serve as flagship projects able to illustrate world-class aspirations and capacity for innovation. While the concept of eco-cities may align with the aspirations of the state, their implementation is tightly controlled by local officials and politicians at the national level.

The conceptual framework for the creation of eco-cities, harmonious society and ecological civilization, can be understood as a political construct attempting to bring together the social concepts of the harmonious society along with the strong environmental ideas associated with low-carbon development. Eco-cities in this context can be understood as a malleable idea that can be shaped to meet differing concepts about ecology and civilization. This is applied in different ways by different promoters as they have different access to governmental subsidies.

However, the current practices of eco-cities in China, and in particular the unchallenged use of capitalist objectives in the implementation of eco-cities, seem to restrict enormously the ability of the eco-city concepts to become a new, fresh and innovative paradigm for Chinese urbanization. The imprecise definitions of what an eco-city is do not make this task any easier. Diverse, at times disconnected, and often contradictory influences on the notion of the eco-city have been present from the very origin of this idea in the 1980s. Eco-cities in China have become subject to the unpredictable processes associated with Chinese planning practices, no matter how inspiring their plans and representations are. The lack of a precise definition of an eco-city means that its conceptualization and implementation strategies can be moulded according to the needs of the developer or politicians. While in Dongtan it can be seen that the project team was unable to negotiate the many political variables, the case of Tianjin illustrates how the notion of a 'sustainable' city is adapted according to the aspirations and needs of politicians. This, of course, is not a complete sample of the rich diversity of eco-cities in China. Municipality-led projects can be analysed independently as they are driven by different ways of attracting the interest of market actors as well as the politicians at the national level.

To conclude, the lack of a clear definition of an eco-city implies that its conceptualization and how it is – or isn't – implemented can be moulded according to the needs of the developer and, especially, state actors. These motivations, however, are not easy to understand – particularly from a Western planning perspective focused on normative compliance. Asian urbanism is highly political and directly guided by the state; China is no exception. It is no surprise then that while eco-city projects in other geographies have been challenged on the basis of their deliverability or ecological credentials, the Chinese cases end up gravitating to the complexity of local politics. The evolution of the content of Chinese propaganda illustrating the natural and built environment shows that images of architecture and masterplanned spaces have become more and more entwined in the political images of national pride and modernization. Architectural imaginations at city scale, in particular those of eco-cities, have increasingly become aligned with the discourses of the CPC, resulting in the adoption of those images by the state at a local level as a marketing tool and at the national level as both marketing and rhetorical tools – with or without the agreement of planners and designers.

Note

1 Suzhou Industrial Park is a joint venture project that includes investment from the Chinese and Singaporean governments. Its site located west of Shanghai, in Jiangsu province, encompasses the development of an area of 80 square kilometres.

References

Adams, R. 2010. "Longing for a greener present: liberalism and the eco-city." *Radical Philosophy* 163: 2–7.

Baeumler, A., Ijjasz-Vasquez, E. and Mehndiratta, S. 2012. *Sustainable Low-Carbon City Development in China*. Edited by Axel Baeumler, Ede Ijjasz-Vasquez and Shomik Mehndiratta. Washington DC: The World Bank.

Brenhouse, H. 2010. *Plans shrivel for Chinese eco-city*. 24 June. Accessed on 21 December 2011 from http://www.nytimes.com/2010/06/25/business/energy-environment/25iht-rbogdong.html.

Cao, H. 2011. "Moving towards low carbon ecological city: investigate the developmental situation in China's first group of low carbon city." *Remote Sensing, Environment and Transportation Engineering (RSETE), 2011 International Conference on*. Nanjing: IEEE. 487–490.

China.org.cn. 2012. *Constitution of Communist Party of China (Adopted on Nov. 14, 2012)*. Accessed on 23 December 2013 from http://www.china.org.cn/china/18th_cpc_congress/2012-11/16/content_27138030.htm.

Communist Party of China. 2007. *Harmonious Society*. 29 September. Accessed on 23 November 2009 from http://english.people.com.cn/90002/92169/92211/6274603.html.

Connelly, J., Lohry, G., Lu, A., Marion, A. and Springer, C. 2012. *Eco-city Notes*. Accessed on 1 June 2012 from http://ecocitynotes.com/map/.

Datta, A. 2012. "India's ecocity? Environment, urbanisation, and mobility in the making of Lavasa." *Environment and Planning C: Government and Policy* 30: 982–996.

Head, P. 2006. "Creating an Eco-city." *Ingenia* 20: 17–21.

Hesse, J. 2010. *Wikileaks: Richard Branson cable*. Accessed on 14 November 2012 from http://realbusiness.co.uk/archive/wikileaks_richard_branson_cable.

Hsing, Y-T. 2012. *The Great Urban Transformation: Politics of Land and Property in China*. Oxford: Oxford University Press.

Joss, S., Tomozeiu, D. and Cowley, R. 2011. *Eco-Cities – A Global Survey 2011*. London: University of Westminster - International Eco-Cities Initiative.

Leihong, P., Zhuang, G. and Zhang, C. 2011. 把脉中国低碳城市发展 *[Feeling the Pulse of China Low-Carbon Urban Development: Strategies and Methods]*. Beijing: 中国环境科学出版社 [China Environmental Science Press].

Liu, J. 2010. *Low-carbon claims by Chinese cities are misleading, says energy expert*. 4 November. Accessed on 11 November 2010 from http://www.guardian.co.uk/environment/2010/nov/04/china-low-carbon-cities-misleading.

Moore, M. 2008. *China's Dongtan demise is mirrored by bad news for Britain's eco-towns*. 18 October. Accessed on 15 May 2011 from http://www.telegraph.co.uk/news/worldnews/asia/china/3223985/Chinas-Dongtan-demise-is-mirrored-by-bad-news-for-Britains-eco-towns.html.

Ng, G. 2009. "Rumblings in Tianjin eco-city." *The Straits Times*, 27 December: n.i.

PRC State Council Information Office. 2011. "《中国应对气候变化的政策与行动 (2011)》白皮书." *The Central People's Government of the People's Republic of China*. 22 November. Accessed on 27 November 2011 from http://www.gov.cn/ jrzg/2011-11/22/content_2000047.htm.

Ross, L. 1988. *Environmental Policy in China*. Bloomington and Indianapolis: Indiana University Press.

Self, J. 2013. "The best laid plans." *The Architectural Review* CCXXXIV 1402: 103–104.

Shin, H.B. 2011. "Vertical accumulation and accelerated urbanism: the East Asian experience." In *Urban Constellations*, edited by Matthew Gandy, 48–53. Berlin: jovis Verlag GmbH.

SIIC. 2011. *Real Estates – Shanghai Industrial Investment Corporation*. Accessed 15 May 2011 from http://www.siic.com/en/business/business_01_01.htm.

SSTEC. 2010. *Celebrating Eco*. Tianjin: Sino-Singapore Tianjin Eco-City Investement and Development Co. Ltd (SSTEC).

Wu, F. 2012. "China's eco-cities." *Geoforum* 43: 169–171.

Wu, F., Xu, J. and Yeh, G-O. 2007. *Urban Development in Post Reform China: State, Market and Space*. London: Routledge.

Xinhua. 2011. *China aims for "ecological civilization."* 17 July. Accessed on 24 January 2012 from http://news.xinhuanet.com/english2010/china/2011-07/17/c_13990879.htm.

——. 2012. *Promoting ecological progress key to socialism*. Accessed on 22 December 2013 from http://www.china.org.cn/china/18th_cpc_congress/2012-11/18/content_27152687.htm.

Xuhui, Y. 2010. *Chongming Eco-island of the world's first line of Dongtan Eco-City is still far away*. 21 January. Accessed on 28 December 2011 from http://www. p5w.net/news/cjxw/201001/t2789262.htm.

Yang, B. and Dong, K. 2008. "Theories and practices of eco-city planning: a case study on Tianjin Sino-Singapore Eco-city." *China Planning Review* 17(4): 32–39.

Yep, S. 2008. "Planning for eco-cities in China: visions, approaches and challenges." *ISOCARP Paper Platform*. Accessed on 12 November 2010 from http://www. isocarp.net/.

Yu, L. 2011. "Shenzhen, China." *Sustainable Places Research Institute*. June. Accessed on 5 August 2012 from http://www.cardiff.ac.uk/research/resources/ yanzhou-presentation-yu-li.pps.

Zhongjian. 2008. *Sino Shanghai Dongtan Eco-City environmental projects stranded*. 12 November. Accessed on 29 January 2012 from http://blog.ifeng.com/article/ 1844470.html.

SLOW

Towards a decelerated urbanism

Abdul Shaban and Ayona Datta

In June 2015, the CEO of Kochi smart city in India suddenly resigned. This came as a shock to many in the sector, given that the inauguration of the first IT building in Kochi smart city was scheduled in July. As stories began to emerge of a three-million-GBP corruption racket run by the CEO, rumours also circulated that he had been removed by TECOM Dubai holdings (the investor) once this was established (Praveen 2015). The corruption was related to the purchase and use of inferior-quality materials at the price of high-spec building materials quoted in the tenders. As TECOM ordered an immediate audit of its finances in Kochi smart city, high-level ministerial committees and the Smart City Council of India rushed to do damage control. An independent audit report thereafter revealed several inconsistencies in the project, finding that the terms and conditions of the smart city and its claims to the provision of jobs had been diluted.

In itself, this story can be seen as the 'ordinary' story of postcolonial urbanism. It highlights the complex web of inconsistencies, ambiguities, corruption, political power games and 'independent' investigations that in the end have failed to bring about the transparency, efficiency and formalization aspired to in postcolonial modernity. Crucially, this ordinary story underlines the perceived 'slowness' that often leads to the deceleration and stalling of mega-projects. It is in this context that planning and bureaucracy surrounding the impending urban age are subject to the logics of speed. Speeding up urbanization through tropes like smart cities are argued to address the corruption, bureaucracy, inertia, nepotism and general unaccountability of those in power that have characterized postcolonial urbanism so far. Slowness here is framed pejoratively, as a stretching of time made possible both by an illegible state and 'anti-development' activists working outside the limits of law. Crucially, 'slowness' is constructed as a 'handicap' of development, modernity and progress. It is ironical then that smart cities as a national urbanization priority in India claim to break this historical connection between urbanization and its vices, but ultimately succumb to its forces.

In the Introduction to this book, Datta argues for a critical lens of 'speed, time and duration' with which the conceptualization, master planning, production and materialization of new cities across the global south are to be examined. She calls these 'fast cities' since they engage in 'claimsmaking' (Lauermann 2015) to urban futures through the imperatives of speed. They combine rhetorics, imagery and 'futurology' around a crisis of urbanization, migration and climate change to present speed as a way out of crises. And it is by constructing relative binaries of speed between 'fast' and 'slow' urbanization paradigms that new cities claim to 'leapfrog' into sustainable urban futures. In doing so, Datta argues in the Introduction that a critical lens of speed can challenge dominant narratives of neoliberalization and global gentrification in postcolonial urbanism. Each subsequent chapter's examination of urbanization in this book critiques the 're-emergence of the postcolonial state desirous of distinction, differentiation and disentanglement from the "colonial burden" – a reinvention through new utopian imaginings of the city'.

The smart city Kochi story nevertheless shows how the faultlines of speed are written into its own imperatives. Smart cities in India are the new fast cities introduced with a wave of rhetoric around their urgency and efficiency in the global urban age. Speed, however, is the new mode of enunciation of sovereign power, where the 'relationship between popular control of government and private control of means of production, distribution and exchange is a fundamental dichotomy in society that tends to play out in favour of business interest' (Davies 2004, 27). Speed in this context becomes the imperative often at the cost of rights, justice and democracy, as we see in several chapters of this book. Yet speed and accelerated urbanization do not in themselves address the historical relations between power and capital. Indeed, if anything, the smart city Kochi story suggests that speed now reconfigures historical relations between state bureaucracy and nepotism into new relations of venality between the state and private sector. Yet several chapters in this book have highlighted that in this transformed relationship, it is grassroots resistance that is often framed as the 'flipside' of speed since it contradicts what social and political elites see as their 'democratic' rights to capital, mobility and middle-class lifestyles.

On the other hand, it is precisely for the above reasons that socialist or autocratic countries such as China or the UAE are seen as progressive by much of the Indian middle-classes when it comes to 'fast' development. Indeed, the urgency of responding to repeated crises has evoked a critique in the global south that planning needs to become more responsive and therefore faster and proactive. New cities are now subject to the rhetorics of 'streamlining', 'leapfrogging', 'smoothening', 'fast-tracking', 'simplifying' and several other phrases that imply the suspension of a democratic state and the ascendance of a particularly aggressive form of the entrepreneurial state (Mazzucato 2014). Speed, when prioritized over democracy and citizenship, becomes a mode of governmentality, a tool of accumulation by dispossession and of forced displacement. Speed articulates an 'acquiescent citizenship' whereby support rather than critique of fast cities becomes the new civic duty of citizens.

Despite the rise of a new state apparatus around speed, there are simultaneous attempts to depoliticize speed. Typically growth coalitions put their case for fast growth as 'value-free' at a political level, while mobilizing local media interests and backing pro-growth politicians and strategies (Houghton 1996, Logan and Molotoch 1987). Emerging political battles along religious, regional, ethnic, caste and class lines in India (Shaban 2010) and rampaging civil wars in Africa, Afghanistan and other parts of the global south are examples of what fast growth may have contributed to on a global scale. At the core of this new (fast) urbanization lies the primitive accumulation of capital – the assimilation and capture of non-capitalist means of production into capitalist ones (the eviction, encroachments and acquisitions are major forms through which the accumulation is accomplished), as an aspect of the formal imposition of 'regimes of accumulation' by the sovereign state. This fast urbanization produces a subsistence–accumulation duality (Bhattacharya and Sanyal 2011, 42), whereby the accumulation economy breaks the subsistence economy to produce structural disadvantages in marginal livelihoods and ways of life. This is most evident now in the 'land wars' waging across several regions of the global south where the interests of urbanization are deflected by peasant articulations of rights to land and livelihoods. New regimes of speed construct the loss of livelihoods as collateral to the 'public' interests of development and economic prosperity. Peasants are now the new frontiers of fast cities and fast urbanization, because their existence is often seen as a threat to speed (Goldman 2012). This legitimization of the need to remove peasants from what has traditionally been understood as their rightful economic space presents the dialectics of fast cities between urban and rural economies, landscapes and lifestyles and by extension between speed and slowness.

In this book we have analysed the rhetorics, politics and practices that follow the ideologies and moral imperatives of fast cities. As several chapters have argued, the consequences of producing new cities and urbanization using the rhetorics of 'fast-tracking', 'leapfrogging' and several other memes of speed produce a violence of development that bypasses processes of democratic inclusion and citizenship. However, slowing down the process of urbanization does not in itself imply the resurrection of social, spatial and environmental justice. Indeed, much of the legitimization of fast cities, as we have argued in this book, follows a model of 'entrepreneurial urbanization' (Datta 2015) that has bypassed local complexities and specificities of capital, governance and citizenship. As Bhattacharya and Sanyal note, 'this unhinging of the cities from their regional or national economies manifests in the dissociation of the new class of workers engaged in immaterial

production from regional lifestyles and prevalent social modes of reproduction' (2011, 44). They find themselves distant from the terms of engagement
with civil society where the demand and aspirations of the middle-class are
often articulated (Bhattacharya and Sanyal 2011). This surplus population
are then forced into the hugely informalized and segmented labour market, having to negotiate with agents of the state to secure their lives and
livelihoods.

Our purpose in this book therefore has been to show that speed, time
and duration are essential components of a critical urbanism. This does not
imply that new cities that face 'blockades' and manifest slower than others are necessarily democratic. We have argued rather for a more careful
consideration of time as a way to enrich (rather than bypass) processes of
democracy, citizenship, sustainability and belonging in the making of cities. The chapters in this book highlight that while inertia in itself does not
embody any inherent guarantees of equality or inclusion, attention to processes of ensuring justice, rights and democracy rather than efficiency might
necessarily need to slow down the pace of urbanization.

Decelerated urbanism: a series of provocations

> Government is often characterised as being too slow, but speed
> should not be a driver in itself. It could be that we need a form
> of slow government, predicated on a similar idea of slowness that
> underpins the slow-food movement: valuing craft, provenance,
> attention to detail, shared responsibility, while creating a platform
> for dialogue and community through human-centredness. The fast
> 'push-button democracy' might well be the last thing we need.
>
> (Hill 2012)

In light of massive structural shifts across the global south, we need to take
heed of Hill's (2012) statement above. Hill captures the challenges of growing 'too fast' by suggesting that 'speed' as a model of urbanization sacrifices
the democratic processes that are usually relatively slow given their attention to the processes of consultation, deliberation and planning. It is in this
context that we argue for a decelerated urbanism to engage with processes
of democracy and citizenship in planning and governance.

But what are the alternative development regimes that produce just and
democratic futures? Almost 20 years ago, Imbroscio (1998) outlined six
elements of an alternative urban development regime. These are strategies
to (i) increase human capital, (ii) increase community economic stability,
(iii) properly account for development costs and benefits through public
balance sheets, (iv) develop asset specificity, (v) increase economic localism and (vi) develop alternative institutions. Imbroscio's paradigm of
local economic development is still relevant to urban America's problems.

But in the context of the global south, his call to cities to adopt strategies of entrepreneurial mercantilism, community-based economic development and municipal enterprise does not capture the entrepreneurial capacity of the state–private sector alliance. It also does not respond to the increasing transformation of state–citizen relations and therefore of the nature of citizenship in the global south.

In the rest of this concluding chapter, we engage in a series of provocations that imagine a multiplicity of different urban futures to that currently enacted across the global south. In provocation #1, we propose slow and 'sensory' urbanism for ushering in new possibilities of urban citizenship, community and spatial justice. We propose that attention to the rhythms of everyday life in cities through sensory and embodied engagement with urbanization will produce a more equitable distribution of power and resources among the grassroots. Provocation #2 emphasizes slow governance leading to a more humane, differentiated, deliberative, participative and contextualized form of urbanism rather than a 'push-button' democracy. We propose that a transition to slow policy and governance might hold the key to urban futures which are in the long term resilient to future crises. In provocation #3, we argue for the need for democratization to stop 'land wars' which the entrepreneurial state is ushering in global south. We propose that the decommoditization of land and the notion of territorial commons is key to an articulation of a decelerated urbanization. Here current urbanization paradigms need to consider the materialities of social and spatial justice that are embodied in claims to land and livelihoods. Finally, in provocation #4, we call for alternative grassroots utopias that can mobilize and materialize what Lefebvre (2004) calls the 'impossible possible' in future urbanization paradigms.

Provocation #1: slow urbanism

Our first provocation is to counter the praxis of fast urbanism. Removing the prerogative of speed in the management and implementation of urbanization projects opens up new possibilities for urban citizenship. This is not to establish binaries or moral positioning between slow and fast cities, but rather to radically reorganize the notion of time, speed and duration in finding local approaches to design and urban planning. Such attention cannot be paid through acquiescence to the entrepreneurial agenda of the state, but rather through a dialectical relationship between the language and performance of rights and the terms and conditions of citizenship imposed by the entrepreneurial state.

We draw here upon a range of 'slow' movements that emphasize the local, sensory and embodied nature of urbanism. One of these, the slow food movement, is an international NGO that focuses on the sensory and embodied qualities of food as an alternative to globalized fast food systems.

It emerged in 1986 and was initiated by an Italian food writer who was alarmed by the opening of a McDonald's restaurant next to the Piazza di Spagna in Rome. This movement was started to keep local community economies vital. The movement's aim was to protect the 'rights to taste' by protecting traditional food products, promoting the pleasure of eating (including the social sharing of a meal) and promoting traditional agricultural methods and techniques, among other initiatives (Mayer and Knox 2006).

Akin to the slow food movement, 'slow cities' or Cittaslow is an alternative movement promoting local sustainability claims to address larger concerns of spatial justice in cities and regions. Cittaslow draws inspiration from the slow food movement and positions itself explicitly against the 'corporate-centred/mainstream economic development policies' (Mayer and Knox 2006, 322), arguing against the speed and homogenizing qualities of globalization that reproduce a specific template of neoliberal urbanism across the world. It aims for the 'creation of a progressive network of small towns – Slow Cities or Citta Lente – that set out to follow an alternative urban development agenda'. It was established in 1999 by the mayor of four Italian towns (Greve in Chianti, Bra, Orvieto and Positano) and president of Slow Food (Lowry 2011). Cittaslow has 54 certification criteria of which 24 are compulsory. These criteria relate to six spheres, namely (i) environmental policies, (ii) infrastructure policies, (iii) technologies and facilities for environmental quality, (iv) safeguarding autochthonous production, (v) hospitality and (vi) awareness (Lowry 2011, 3). To be certified by Cittaslow, a city must also have a population 50,000 or less, which implies that the potential for transformation lies in small and medium towns. In 2011 there were 141 Cittaslow cities in 23 countries certified by Cittaslow International. Although all of these cities are in Europe, their policies to incorporate slowness and conviviality in urban life can be translated into a diversity of regional and local contexts.

As Pink (2008, 106) notes, the principles of Cittaslow 'engage their participants sensorially rather than simply economically, intellectually or emotionally'. Pink further notes that a 'sensory approach to urbanism can produce insights that contribute, alongside conventional methodologies, to our understandings of human engagements in sustainable urban development processes'. Cittaslow emphasizes local economic strengths that contribute to equality and community mobility. Mayer and Knox further argue, 'Slow cities are places where citizens and local leaders pay attention to local history and utilize the distinctive local context to develop in better and more sustainable ways' (2006, 322). But Cittaslow, like several other movements around local sustainability, is not without its weaknesses. Its localization of policy and governance embodies inherent assumptions that have been the subject of continued debate in recent urban studies scholarship. The assumption that the local is somehow more equal and equitable than the regional or national assumes that networks and structures of power and capital do not work within and across local communities. It also assumes that devolution of political power to the 'local' (including urban local bodies and local authorities) encouraged by international development agencies in global south countries will automatically ensure equitable distribution of capital and resources and decision-making across socially and economically marginalized groups. While several countries in the global south have initiated policies of Agenda 21 to decentralize and localize decision-making powers, the emphasis on localism often assumes that greater freedom or 'spatial liberalism' (Clarke and Cochrane 2014) will promote innovation and greater engagement among civil society. In countries such as India and China, this localism has instead consolidated the power of local elites and middle-classes while further pushing the marginalized to the peripheries. Moreover, as many of the chapters in this book point out, localism can itself become a form of 'statecraft' whereby power, inclusion and participation in local decision-making are mediated and controlled by the state.

A slow urbanism in the global south would move beyond a mere recognition of local history and contexts in the manner advocated by Cittaslow or other 'slow' movements to a historiography of social inequalities, accumulation, dispossession and exclusion. Slow urbanism would articulate an attention to the prosaic transactions of everyday life in cities, to the emotions, feelings and experiences of those who inhabit the city, of those who are directly affected by the economic imperatives of urbanization and fast-paced growth. This means a thick description of the ways in which social inequalities have been historically connected to urban development, a critical analysis of the contexts in which these social inequalities have been understood through linear time, and a study of the ways in which the notion of cyclical time has been devalued in critical urban geography, often to the detriment of subaltern marginalized groups. In short, slow urbanism will not just acknowledge the spatio-temporal cycle of inequalities and exclusions, but will engage, debate and seek to address these repeatedly during the cycles of urbanization.

What would a slow urbanism look like? In the first instance it would take heed of cyclical time. In the Introduction to this volume, Datta draws upon Lefebvre's notion of rhythm analysis, which 'concerns the everyday, rites, ceremonies, fetes, rules and laws, there is always something new and unforeseen that introduces itself into repetitive difference' (Lefebvre 2004, 6). This is fundamentally in opposition to the shaping of fast cities through the marking of linear time in global urban crises of migration, urbanization and climate change. Slow urbanism takes heed of what Lefebvre notes as 'the lived, the carnal, the body' (2004, 9) in processes and policies of urbanization and would lay the foundations of future urbanization on social justice and citizenship. Lefebvre notes, 'we know that a rhythm is slow or lively only in relation to other rhythms (often our own: those of our walking, our breathing, our heart)' (Lefebvre 2004, 10). We can make these rhythms feature in urbanization by considering how the body occupies the processes and practices of urbanization, not just in being present in new cities, but also by being rendered invisible by fast urbanism. We can further examine the dialectical relation between the notion of linear time in the fast construction of new cities and the notion of cyclical time in the ghost cities left behind after the construction boom. Slow urbanism would allow for a more nuanced connection between the rhythms of global urban crises, economic downturns and migration explosion and the cyclical time of exclusions, dispossessions and expulsions. A slow urbanism would prioritize the time needed in the making of ordinary cities rather than the speed of making fast cities.

Provocation #2: slow governance

Our critique of fast cities in this book has sought to argue that speed embodies the logics of capital accumulation and growth, which are engaged in the governance of 'population' rather than the advancement of diverse citizenships. This fast urbanism is promoted through the assumption that speed (aka efficiency) can achieve 'good governance'.

Fast cities are a 'gold rush' for the 'experts' (Bhatia 2016) which has led not only to a north–south transfer of models of urban governmentality but also increasingly to south–south collaborations and partnerships between nation states and urban municipalities. All the cities studied in this volume have significant involvement from experts such as IT consultants, architects, planners and state officials in the north and south who are continuously looking for avenues to further their growth through consultancy, providing expert advice and exporting their people and technologies to market new 'governance models' in the global south. This fast governance market assumes uniform conditions of economy, ecology, polity and social behaviours. A good example of this is the recent national urbanization drive in India to create 100 smart cities which has become an archetype of 'fast policies' for 'fast cities', which advocate information technology-driven solutions to all problems from traffic to security and e-governance.

In this drive, fast governance through different forms of communication systems, new media technologies and ICT has become a euphemism for democratic participation and citizenship. The faster the channels of governance, it is argued, the more efficient and transparent the system of governance. To this end, we have seen in this volume how nation states have begun to radically change the terms and conditions of citizenship, territorial belonging and democratic participation through the rule of law and judiciary. Efficiency is now a cul-de-sac of governance where often the curation of Facebook likes, shares and Twitter retweets, or what Dan Hill calls 'push-button' democracy, has begun to stand for deliberative and participatory models of governance. Indeed in China, India and other global south countries, citizenship has increasingly begun to mean compliance and participation in state entrepreneurial models of policy and governance.

It is not surprising then that fast governance remains geared towards restricting dissent, since dissent inherently 'slows' down its pace. This extends to areas of public disclosure law, rights to information, community participation in planning processes and land acquisition, as shown in several chapters of this book. Yet exclusion, polarization (economic and political), disarticulation (Banerjee-Guha 2009) and 'protestations' remain major characteristics of mega-urbanization and city making in the global south today. The present paradigm of fast policy led by a global neoliberalism garners increasing support from the middle-classes, weakening grassroots movements and often perpetuating plutocracy and majoritarianism in the name of democracy (CASUMM 2006). The poor are reduced to 'vote banks', and their interests are pejoratively interpreted. This in turn creates 'shareholder democracy' (Durington 2011, 209) through industrial, house-owner and shareowner associations enjoying the 'club good' benefits (Atkinson and Blandy 2006) that fundamentally violate basic (social) democratic principles. While technology governance is now an integral part of state governance, its claims to accelerated results and efficiency are highly overstated and misleading. Indeed, fast governance has begun to stand for a governance by exception.

It is in this context that we need an alternative and radically different model of slow governance. Slow governance should not be misinterpreted as a return to the model of nepotism and micro-bureaucracy, but rather as a considerable slowing down of the speed of urbanization that will provide opportunities to apply 'slow thinking' (Kahneman 2002) and a reasoned approach to context. Slow governance would allow for innovation and entrepreneurialism in ways that enrich rather than erase the potentialities of a variety of citizenship practices and performances. But it would do this without the need for embedding citizenship within the spaces of corporate-run e-governance sites such as those owned by Facebook, Google or Twitter. It would instead enrich and encourage a diversity of citizenships from the intimate to the global across various spaces and times. We elaborate further on this in the next section.

Provocation #3: democratize the commons

There is a side to the 'speed' with which urbanization has been conceived and executed in the global south that has not yet been explicitly connected to the land question. Examining the 'land wars' between the private investors and farmers in India, Levien (2012, 8) notes that the Indian state appears to be caught between the land requirements of its liberalized growth model and the exigencies of electoral democracy. This observation is significant since the Indian government as a representative democracy is elected by a population that is still largely rural and yet is driven by the aspirations of a rising urban middle-class who desire increased global investment into its cities and regions to benefit from its prosperity. It is therefore stuck in a conundrum – its grandiose plans of attracting investment to speed up urbanization regularly collide with the more prosaic realities of making land available for this investment to materialize.

One of the key aspects of a decelerating urbanism and slow governance would be to overturn the notion that land as a form of livelihood or commons can only become productive if converted to real estate. In other words, the fairy-tale of fast urbanization makes us believe that development can only occur when governance of land is tied to urbanization. In countries such as India and China, real estate remains the highest-growing sector, where the capital from other sectors is invested due to high speculative returns. The speculation, monetization and marketization of land require property rights which are often acquired from peasants at a cheap rate or transferred by governments to individuals by acquiring commons or public land. As Schindler (2015) has suggested, countries in the global south are now more interested in the governance of territories through which they aim to govern populations. In a country like India, land acquisition by the state and corporate sector in the last two decades or so has emerged as the priority of national development policy. It is assumed that land transfer to global enterprises by the entrepreneurial state will create significant benefits to the people and aid economic growth. Studies have shown that neither urban centres nor the industrial sector for which major chunks of land have been acquired have helped in trickling down the benefits or creating employment for the people (Bhattacharya and Sakthivel 2004). In fact, land acquisition policy has created new forms of inequality that have changed the basic security system of farmers, tribals and other land-dependent communities. In the scheme of cities as business models, the strategy of public land acquisition has largely been of a transfer of wealth from poorer sections to the corporates and real estate developers.

The experience in several countries, as evident from the chapters in this volume, suggests that the urbanization question is really a land question. As Shin shows in Chapter 5, the building of Songdo City in South Korea is based on the promise of a real estate boom. In fact, Songdo City has seen the first foreign ownership of Korean soil where a parcel of reclaimed land was sold to a joint venture whose majority share was held by a US real estate developer. Thus while urbanization might be measured in economic growth rates, its continuous demand for land as cheap raw material highlights the terrain of governmentality beyond the normative questions of technical implementation and efficiency.

What would cities look like if real estate was not seen as an indicator of growth under neoliberal urbanism? Would land still hold the same value for urbanization? Would land still need to be violently 'tamed' to serve the purposes of capital?

Sampat (2015) argues for 'a legal framework for land- and resource-use that is locally determined, egalitarian and ecologically appropriate as a tool towards ushering a fundamental reconstitution from below'. Democratizing the urban and rural land as commons will provide the opportunity to grass-roots and marginalized communities to produce a slow economy that will be more sustainable in the long term. This includes both a legal framework for democratization that will not criminalize dissent and rights claims, and also a cultural transformation in the understanding of private property. Models of such democratization already exist in the form of Commons Property Management (CPR) more recently and the 'Bhoodan' (land gift) and 'Gramdan' (village gift) movements in early 20th-century India. These are also present in the form of Community Land Trusts in the USA. These models themselves are not without their challenges, but it is worth discussing them briefly to explore their potential in achieving democratic resource use.

Bhoodan and Gramdan movements in India emerged as a Gandhian ideology of 'sarvodaya' (or development for all). Under this movement, the gifting of land (bhoodan) was seen as a non-violent process of land distribution to those at the bottom of the social hierarchy. Bhoodan was prevalent in India in the 1950s–60s and land transfers under this model were legalized by the state. It was also a socialist experiment based on the activism of prominent Gandhian ideologue Vinoba Bhave, and in that sense was directed towards social justice and grassroots democracy. Gramdan fol-lowed in the years after Bhoodan where land from the entire village was handed over to the village community, making every villager a stakeholder in the common property. The produce and earnings from the land were also shared collectively to benefit all villagers. Land could not be sold or acquired without permission from the village council. Although these two movements had several weaknesses (related to records of land distribution and failure to transform historical social power) and faced several challenges in implemen-tation (corruption, take-up and commitment from the collective) at larger scales, their ideology was translated and institutionalized in the USA as Community Land Trusts. Community Land Trusts (CLTs), which continue to this day, are non-profit organizations which maintain collective access to land, property, housing and so on for communities to provide affordable, equitable and sustainable access to resources.

What shape and form could such institutions of democratic common-ing take in the global urban age? What future do these institutions have in a global neoliberal economy that drives fast urbanism in the global south? On the one hand, if these institutions work to their potential, they will provide blockades to the velocity of contemporary global urbanism. On the other hand, we cannot have a return of the same institutions given that the social, cultural, legal and political contexts in which they arose are now transformed. The postcolonial nation-state, in its role as neoliberal entrepreneur, now needs more than ever before 'to unlock land values' to maintain its sovereignty and rule over territories. This 'need', however, is sugar-coated in the language of public prosperity, growth and development. Indeed, recent institutions such as the Gramsabha (village council) in India have often worked as collaborators of private capital to the detriment of landless farmers. Using their powers, the Gramsabha has tended to prioritize capital accumulation by sanctioning land use change from agricultural to industrial purposes. This has shifted power and resources further into the hands of the landholding families, while dispossessing landless farmers. The reverse is seen in the Chinese Hukou system, where national household registers fixing populations to territories restrict rural migrants from accessing benefits in the urban commons. While they work and live in urban areas, they are unable to access affordable housing or bring their families to stay with them in the city.

We still do not have all the necessary tools to understand the complexities and consequences of democratizing the commons. But what is clear is that any gesture in this direction will need to be backed up by strong policy and implementation. This necessarily means slower processes of collectivization and activism to agree and demarcate commons property around resources such as land, water bodies and forests. This also means finding alternatives for growth that are not rooted in the need for manipulating territory. It means finding new ways of urbanizing, new ways of governing and new ways of deliberating upon the encroachment from fast cities onto the commons.

Provocation #4: alternative utopias

[I]n order to extend the possible, it is necessary to proclaim and desire the impossible. Action and strategy consist in making possible tomorrow what is impossible today.

Henri Lefebvre (1976, 36)

The above provocation by Lefebvre is counter to the current urbanization model. Each of our provocations above follow this countercurrent by demanding a radical reorganization of the very structures of the state and neoliberal urbanization. We hope to have shown that these are viable imaginations of more just futures, indeed of possible futures. The examples we have used certainly suggest that the impossible can be made possible as long as we are able to imagine these. Indeed, urban utopias are subjective. In other words, what seems impossible to some might be mundane and everyday to others.

In his work of creative non-fiction, Darran Anderson (2015) notes in *Imaginary Cities* that a successful utopia will be a plurality. Moving away from the corporate-driven imagination of utopia (in the form of big data, smart cities and Internet of Things), Anderson notes that what was understood as utopia in the past is already here for many urban citizens in the form of piped water, electricity, jet planes, global communications and so on. The future and therefore utopian futures were all once utopian. When utopias are made possible, they become invisible, but they also exist in fragments. Crucially, he notes that it is only when they are taken away that we realize the utopian aspect of cities. He suggests that we need to define utopias in advance, to locate where the impossible possible does exist, to highlight them and critically emulate them elsewhere. It is only in rethinking how we understand the past and the future that we can imagine a possible utopian future that is radical and socially just.

Anderson's (2015) notion of utopia presents several important questions. Where does the future begin and end when we are talking about utopias? Clearly if utopias are subjective, fragmented and spatio-temporal, then our notions of utopia need to change across time, and most importantly our assumptions of utopian impossibility need rethinking. How does the notion of time wrapped up in the future reconstitute our very ideas of utopian thinking? How do we spatialize time, speed and duration when we imagine possible utopian futures?

In the Introduction to this book, Datta noted that speed is the new urban utopia. This notion of utopia refers to the forms of thinking that produced architectures and built environments of grandiose claims in the modernist period. This form of utopian thinking has received several elegies to its 'paradoxical call to order, an atavistic alliance with modernist dreams of total environmental control' (Martin 2010, 1). Indeed, failed modernist utopias can be seen as the collective failure of the state and of the built environment professions of architecture and planning. The new utopias of fast cities now present new collective imaginations of state and corporate sector alliances. Social order in these new utopias is now largely vested in and through ICT architectures, where digitally directed planning and urban life is the imagination of the future. Digital communications, big data and Internet of Things are now ways to speed up fragmentary and uneven access to a previously unimagined future of ubiquitous connectivity.

In this context of a neoliberal capture of utopian thought and praxis, what other alternative utopias can be imagined? How can utopian thinking move beyond the representation of efficiency, immediacy, convenience and ubiquity (Anttiroiko 2013, Hollands 2008, Johnson 2013) to political stakes in rights, justice and citizenship? David Harvey (2000) in his book *Spaces of Hope* provides several provocations to think about more hopeful urban futures. He argues that the purpose of utopia is not to 'provide a blueprint for some future but to hold up for inspection the ridiculous foolishness and waste of the times, to insist that things could and must be better' (1998, 281). He advocates a 'dialectical utopianism' (240) which means a radical project where the rule-making nature of communities can confront the rule-breaking nature of insurgent politics. For Harvey (2000), this dialectical utopia is not just spatial or temporal. Rather it is 'spatio-temporal', which means it requires the continuous rearrangement of the nature and purpose of imagined politics in different spaces, places and times. This involves experimentation with different material forms of utopian imaginations, 'to explore the wide range of human potentialities' (77) that must materialize to provide radical alternatives to authority and power. Where does that leave the possibilities of slow urbanism in the global south?

In our final provocation we call for a utopian imagination of the 'impossible possible' (Lefebvre 1976) to emerge. This utopia emerges from outside the trappings of a global neoliberal urbanism or a postcolonial entrepreneurial state which have so far captured the rhetorics and practices of alternative urban futures. Its utopian praxis is reimagined as a radical alternative to the violence of speed as a logics of social order and control. This form of utopia imagines not just radically alternative urban futures but also radically alternative citizenships that are not held hostage to neoliberal compliance.

Pinder (2005) calls this critical utopianism. For Pinder this is a 'partisan of possibilities' which needs to break away from traditional, top-down 'abstract ideals and formal plans' (245) of blueprint utopias. It follows claims of Marxist scholars such as Lefebvre and Harvey in arguing that utopianism is a prerequisite for the imagination of emancipatory futures. Closing down or rejecting utopianism based on its historical attempts at concretizing totalitarian social order is a closing down of debates on the possibilities of changes in social and political life. Following Lefebvre and Harvey, Pinder notes that a critical utopianism recognizes the diversity of human experience and raises questions on the role of power in the making of cities and citizenships. Thus utopian thought and action are processes continually in the making and imagining of just urban futures by calling for a transformation of everyday life along the lines of justice, rights and equity. It also means that utopian thought can reveal the inherent injustices, repressions and violence embodied by the urbanization of the state and citizenship even as it seeks to explore and imagine new possibilities and alternatives.

Such a critical utopianism is inherently 'slow'. It is also ideologically opposed to the traditional format of utopian social engineering through the built environment. This utopianism is dialectic and spatio-temporal and therefore directly relevant to the possibilities of a postcolonial critical urbanism. It is iterative and fluid which is informed by critical reflection on spaces, bodies and power. It sees the possibility of emancipatory futures only as the continual imagination of alternatives to fast urbanization. Slow urbanism is a radical alternative utopia which aims to make the impossible possible, while resisting the idea that slow cities are the final outcome of critical utopianism. Slow urbanism is an embodied politics at several scales which conceives of a future that is open to impossible possibilities.

As Doreen Massey would say, 'only if we conceive of the future as open can we seriously accept or engage in any genuine notion of politics. Only if the future is open is there any ground for a politics which can make a difference' (Massey 2005, 11).

References

Anderson, D. 2015. *Imaginary Cities*. London: Influx Press.

Antitiroiko, A-V. 2013. "AI & Society." *Knowledge, Culture, Communication* 28: 491–507.

Atkinson, R. and Blandy, S. 2006. *Gated Communities*. New York and London: Routledge.

Banerjee-Guha, S. 2008. "Space relation of capital and significance of new economic enclaves: SEZs in India." *Economic and Political Weekly* 33(47): 51–59.

Bhatia, R. 2016. "We don't need IT here: the inside story of India's smart city gold rush." *The Guardian*, 22 January. Accessed on 24 January 2016 from https://www.youtube.com/watch?v=zAB5AC9yhY0.

Bhattacharya, B.B. and Sakthivel, S. 2004. *Economic Reforms and Jobless Growth in India in the 1990s*. Working Paper Number 245, Institute of Economic Growth, New Delhi, India: University of Delhi.

Bhattacharya, R. and Sanyal, K. 2011. "Bypassing the squalor: new towns, immaterial labour and exclusion in post-colonial urbanization." *Economic and Political Weekly* 46(31): 41–48.

CASUMM. 2006. *JNNURM: A Blueprint for Unconstitutional, Undemocratic Governance*. Collaborative for the Advancement of the Study of Urbanism through Mixed Media.

Cittaslow International. 2011. "Cittaslow." Accessed on 28 July 2016 from http://cittaslow.org/download/DocumentiUfficiali/CITTASLOW_LIST_2011.pdf.

Datta, A. 2015. "New urban utopias of postcolonial India: entrepreneurial urbanization in Dholera smart city, Gujarat [Anchor paper]." *Dialogues in Human Geography* 5(1): 3–22.

Davies, J. 2004. "Can't hedgehogs be foxes, too? Reply to Chance N. Stone." *Journal of Urban Affairs* 26(1): 27–33.

Dimanti, D., Borelli, N. and Bernadi, M. 2014. "Smart and slow city: the case study of Milan Expo 2015." *Arte-Polis 5 International Conference – Reflections on Creativity: Public Engagement and the Making of Place*.

Durington, M. 2011. "Review essay – gated communities: perspectives on privatized spaces." *International Journal of Urban and Regional Research* 35(1): 207–210.

Goldman, M. 2011. "Speculative urbanism and the making of the next world city." *International Journal of Urban and Regional Research* 35(3): 555–581.

Harvey, D. 2000. *Spaces of Hope*. Berkeley, CA: University of California Press.

Hill, D. 2012. *Dark Matter and Trojan Horses: A Strategic Design Vocabulary*. Moscow: Strelka Press.

Hollands, R. 2008. "Will the real smart city please stand up?" *City: Analysis of Urban Trends, Culture, Theory, Policy, Action* 12(3): 303–320.

Houghton, G. 1996. "Local leadership and economic regeneration in Leeds." In *Corporate City? Partnership, Participation and Partition in Urban Development in Leeds,* by G. Houghton and Collins C. Williams (eds.), 19–40. Aldershot, UK: Avebury.

Imbroscio, D. 1998. "Reformulating urban regime theory: the division of labor between state and market reconsidered." *Journal of Urban Affairs* 20(3): 233–248.

Johnson, L. 2013. "Petropolis now." *New Statesman,* November: 8–14.

Kahneman, D. 2001. *Thinking, Fast and Slow.* New York: Farrar, Straus and Giroux.

Lauermann, J. 2015. "The city as developmental justification: claimsmaking on the urban through strategic planning." *Urban Geography,* DOI:10.1080/02723638 .2015.1055924

Lefebvre, H. 1976. *The Survival of Capitalism: Reproduction of the Relations of Production,* translated by Frank Bryant. New York: St. Martin's Press.

——. 2004. *Rhythm Analysis.* Paris: Continuum.

Levin, M. 2012. "The politics of dispossession: theorizing India's global land wars." *International Conference on Global Land Grabbing II, October 1–19.* Ithaca, NY: Department of Development Sociology, Cornell University.

Logan, J. and Molotach, H. 1987. *Urban Fortunes: The Political Economy of Place.* Berkeley, CA: University of California Press.

Lowry, L.L. 2011. "CittaSlow, slow cities, slow food: searching for model of development of slow tourism." *Seeing the Forest and Trees – Big Picture Research in a Detail Driven World.* Trade and Tourism Research Association. 19–21.

Martin, R. 2010. *Utopia's Ghost: Architecture and Postmodernism, Again.* Minneapolis: University of Minnesota Press.

Massey, D. 2005. *For Space.* London: Sage Publications.

Mayer, H. and Knox, P. 2006. "Slow cities: sustainable places in fast world." *Journal of Urban Affairs* 28(4): 321–334.

Mazzucato, M. 2013. *The Entrepreneurial State: Debunking Public vs. Private Sector Myths.* London: Anthem Press.

Pinder, D. 2005. "Reconstituting the possible: Lefebvre, utopias and the urban question." *International Journal of Urban and Regional Research,* DOI: 10.1111/1468-2427.12083

Pink, S. 2008. "Sense and sustainability: the case of the slow city movement." *Local Environment* 13(2): 95–106.

Praveen, M.P. 2015. SmartCity takes a credibility dip. *The Hindu.* Accessed on 28 July 2016 from http://www.thehindu.com/news/cities/Kochi/smartcity-takes-a-credibility-dip/article7722300.ece.

Sampat, P. 2015. "Right to land and the rule of law: infrastructure, urbanization and resistance in India." *CUNY Academic Works.* Accessed on 28 July 2016 from http://academicworks.cuny.edu/gc_etds/619.

Schindler, S. 2015. "Governing the twenty-first century metropolis and transforming territory." *Territory, Politics, Governance* 3(1): 7–26.

Shaban, A. 2010. *Mumbai: Political Economy of Crime and Space.* Hyderabad, India: Orient Blackswan.

INDEX

INDEX

17–18, 21, 32, 59–60, 93, 124, 126, 131, 135, 137; wars 207, 209, 213
Lefebvre, H. 3, 4, 11, 22, 165, 209, 211, 216, 218

master plan/s 18, 54–5, 57, 59, 61, 125, 136–7; planning 18–19, 57–8, 69, 89, 182, 184–5, 206
mega-city/ies 35, 40, 151; *see also* mega-projects
mega-project/s 16, 33–8, 41–3, 46–7, 83, 85, 102–4, 106–14, 117–18; planning of 169, 176, 183
middle-class 7–8, 22, 54, 60–4, 95, 124, 126–7, 136, 143, 152, 158, 190, 206, 208, 211, 213–14; *see also* civil society
modernism: technological 10; urban 57–9
modernist 2, 18, 32, 35, 217; development 9; planning 10, 11, 36–7, 57, 58; utopia 10, 34, 217

neoliberal: agenda 174, 178; era 17; governance 37; homegrown neoliberalism 125; logics of capital accumulation 84, 94; policies 95; state 14; urban development 7, 13; urbanism 3, 4, 9, 210, 214, 218; urbanization 84, 138, 216
new cities 1–10, 12–16, 18–19, 22–3, 31–2, 35–6, 59, 62–3, 67, 79, 176, 189, 206–8, 212; *see also* new towns/hips
new towns/hips 6, 9, 15, 17, 20, 91, 123–33, 136–40, 142–4, 152, 158, 160, 188, 200; *see also* new cities

parallel cities 12, 31, 32, 35–7; *see also* new cities
petro-: politics 16, 109, 119; urbanism: 16, 42, 108–10
Pinder, D. 11, 218
planetary urbanization 66; *see also* urbanization
policy mobility 4–5, 173
post-carbon economies 16
postcolonial: nation-state 216; subjectivities 19; urbanism 4–6, 8, 14, 23, 205–6
private: cities 7–8, 38; urbanism 8

privatopolis 7, 89
public-private partnership 2, 16, 37, 42–3, 84, 93, 127, 136, 163, 166

Qatar 16; geopolitical ambitions 105; global events 105; Knowledge City 2; Qatar Foundation 103, 113–22, 118; role of the elites in 109, 118; sovereign wealth fund 104–5; urbanization in 108

Rajarhat (New Town): 2, 17, 123–47
real-estate utopia 16; *see also* utopia
regime of accumulation 15, 124
representation: architectural 189, 195; messages of 21; visual power of 8, 19, 194
representational: work 19; power of 21
re-urbanism 32–4, 47; *see also* urbanism
rogue urbanism 2, 11; *see also* urbanism

satellite cities 33, 37, 39–41, 46–8, 60–3, 171; *see also* cities
Shanghai (China) 40, 60, 89, 189, 192–4, 196–9, 203; *see also* Chinese cities
Singapore 40, 42, 60, 79, 87, 91, 109–11, 189–90, 192–3, 196–7, 199–201, 203; *see also* East Asia
slow: cities 23, 210; democracy 13; food 210; food movement 209; governance 209, 212, 213; policy 209; urbanism 211–12, 218
smart city 11, 83, 89, 91–2, 205; *see also* utopia
Songdo 16, 35, 83–95; *see also* Korea, South
South Africa 2, 33, 44–5, 48, 172, 181–2; *see also* African cities
Special Economic Zone (SEZ) 20, 35, 39, 124, 157–8, 163, 202
speed 2–8, 11–13, 16–17, 19–23, 34–6, 66–8, 70, 74, 76, 78, 124–5, 143, 155, 184, 205–10, 212–14, 217–18
sustainability 12, 19–21, 32, 60, 66–71, 75–6, 78–9, 96, 158, 169, 171–3, 175–85, 189–91, 194, 198–201, 208, 210

223

Milton Keynes UK
Ingram Content Group UK Ltd.
UKHW040103071024
449327UK00019B/763